Video Tape
Recorders

SECOND EDITION

by

Harry Kybett

Howard W. Sams & Co., Inc.
4300 WEST 62ND ST. INDIANAPOLIS, INDIANA 46268 USA

Preface

This book is aimed at the user of the helical vtr in education, entertainment, industry, and broadcasting. It is not a book of electronic circuit theory nor of studio and production techniques, since this material is available elsewhere. Instead, it explains in simple language and diagrams how the helical vtr works.

There has been a great lack of published material covering the helical vtr. Most of the engineers involved in the development of video tape recorders have worked for one or, at most, two manufacturers, and therefore their knowledge has been limited to relatively few of the available models. Since most of these engineers have not been disposed to writing, almost all of the literature has been in the form of service manuals, which by their nature are limited to a specific model and are not books of basic theory.

The author has been fortunate in that he has used and serviced nearly all the major models over a period of years and consequently has acquired a broad background of experience. Thus the contents of this book are the result of actual experience rather than literature research. However, much of the information is from service manuals, and much has been gained from discussions with members of the engineering staffs of various manufacturers, all of whom have been courteous and helpful.

Since the first edition of this book was published, two major technological developments have occurred in the tv industry. The first is the successful digitizing of the tv signal, and the appearance of commercial products using these new techniques. The most important of these are the time-base corrector (TBC) and the digital standards converter. The second major advance has been in the mechanical construction of the helical vtr. This has been improved to the point that when a helical machine is used with a TBC, it has a performance comparable to that of the broadcast quad machine.

The result of these developments is that the situation regarding helical vtr's has changed completely. A number of TBCs are available, and several manufacturers now produce broadcast-quality helical vtr's. Together, these devices are now used seriously in network tv studios, teleproduction studios, and the larger industrial and educational studios.

To reflect these changes, new chapters have been inserted to describe the basics of these new devices. To make room for the new material, some material on the older and obsolete smaller helical machines has been dropped.

This book will provide both those who use and those who service helical vtr's with a better understanding of vtr principles, circuitry, and mechanics. It divides the helical vtr into well-defined sections, and then examines and explains the reasoning behind the choices made and shows several alternative methods of achieving the desired ends. All of the information is taken from current models made by several manufacturers, and represents a good cross section of current practice in the industry. As time passes, the circuit details of the various models may change, just as with any electronic equipment. But the basic principles and the mechanics are well established, and the descriptions given can be relied upon to last for a long time.

I am indebted to the following manufacturers for permission to use their names and sections of their service manuals, and for supplying photographs and drawings: Akai America, Ltd.; Ampex Corporation; Robert Bosch Corporation; Echo Science Corporation; Electronic Engineering Company of California; Hitachi Shibaden Corporation of America; International Video Corporation; Panasonic Company; Sanyo Electric Inc.; Sony Corporation of America; Spin Physics, Inc.; and Thalner Electronic Laboratories, Inc. In particular, I would like to express my thanks to Bob Thalner of Thalner Electronic Laboratories, Inc., who allowed me access to his laboratory and files, thus making this book much easier to write.

HARRY KYBETT

CHAPTER 5

THE HELICAL VTR 51
Fundamentals of the Helical VTR—Slant-Track Principles—Head and Drum Arrangements—Tape Guides—The Tape Deck—The Capstan Assembly—Electronics—Conclusion

CHAPTER 6

THE MECHANICS OF HELICAL VTRs 76
Head-Drum Assembly—Smooth Surfaces—The Head Drum and the Recorded Tracks—Tape Formats—Heads—Head Drive—Head Degaussing—Head Connections—Head Changing—The Capstan Assembly—Tape Tension—Motors—Belt Drive—Rim Drive—Reel Tables——Mechanical Linkages and Levers—The Other Heads

CHAPTER 7

THE RECORD AND PLAYBACK SYSTEM 110
Video Recording—Video Playback—Electronics to Electronics

CHAPTER 8

SERVOS 138
Servo Fundamentals—Types of Servo in Common Use—Input Signals—Comparators—Control Circuits—Rotating-Mechanism Drivers—Rotating Mechanisms and Control Devices—The Use of Multivibrators in Servo Circuits—Special Integrated Circuits——The Head Servo—Record Mode—Playback Mode—Ampex 7500—Sony EV 320 Series—Sony AV 3600—Shibaden SV 700—Sony AV 3650——The Capstan Servo—Sony EV 320F—Sony AV 3650——Conclusion

CHAPTER 9

QUAD SERVOS 185
Head and Capstan Motor Drives—The Head Servo—The Capstan Servo—Servos in the Playback Mode—Servo Comparators—Summary

Contents

CHAPTER 1

A Brief History of Video Recording 9

CHAPTER 2

Review of Audio Recording 12
The Principle of the Tape Recorder—Recording a Signal Onto Tape
—Heads—Recording Bias—Playback—Losses—Equalization—Audio
Tape Recorders—Conclusion

CHAPTER 3

Principles of Video Recording 26
Frequency Range Limitations—The Method Adopted—Servos—Head
Switching—Tape Tension—Space Loss—Conclusion

CHAPTER 4

The Broadcast Quad-Head Recorder 39
The Head Assembly—The Tape Path—Record Mode—Playback
Mode—The Quad-Head Servos—Head Switching—Quadrature Error
—Other Errors—Head Wear—Conclusion

CHAPTER 10

CONTROL PULSES AND OTHER FUNCTIONS 199
    Dropout Period—Head Switching—Dropout Compensation—Slow
    Speed and Still Frame—Meters

CHAPTER 11

EDITING 212
    Servos for Editing—Editing Erase—The Control Track—Record and
    Erase Current Switching—Assembly Editing—Insert Editing—The
    Stop Button—Audio—Examples of Editing Facilities—Editing Ac-
    cessories—Mechanical Splicing

CHAPTER 12

COLOR RECORDING AND PLAYBACK 236
    The Direct Method—Color Playback Problems—Direct Record Color
    Correction—The Down-Converted Subcarrier Method—Color Cor-
    rection—Circuits for the Converted-Subcarrier Method—Color Cor-
    rection Review—General Topics Related to Color

CHAPTER 13

TIME-BASE ERRORS AND THEIR CORRECTION 266
    The Main Causes of Time-Base Errors—Quad Timing Problems—
    Helical Time-Base Errors—The Digital Time-Base Corrector—
    Conclusion

CHAPTER 14

VIDEO CASSETTE MACHINES 283
    Cassette Machines—The Cassette—The Mechanics of Cassette Ma-
    chines—Head Drum—Electronics—Operation of the Machine—The
    Smaller Cassette Machines

CHAPTER 15

BROADCAST HELICAL VTRs 305
  Broadcast VTR Requirements and Formats—The Requirements for a
  Broadcast Helical VTR—The Hybrid, or Segmented, Helical VTR—
  Nonsegmented Broadcast Helical VTRs —— *The Machines* — The
  Ampex VPR-2—The Sony Machine

CHAPTER 16

THE PORTABLE VTR 364
  The Portable Camera—The Portable Deck—The Servo in Recording
  —Playback Servo Operation—The Control Track—The Color Portable
  —Portable Cassette Machines—Portable Broadcast Helical Machines
  —Conclusion

CHAPTER 17

INTRODUCTION TO DIGITAL TV 383
  Digitizing an Analog Signal—Analog-to-Digital Conversion—Digital-
  to-Analog Conversion—Sources of Error—The Video Signal—Video
  A/D Converter—Applications

INDEX 395

1

A Brief History of Video Recording

With the development in the early 1930s of an electronic tv signal and the audio tape recorder, and their coming into broadcast use in the mid 1940s, it was not long before serious consideration was given to putting a tv signal onto tape. It had always been possible to record a tv picture by taking a movie of the kinescope screen, but even though such pictures were used for rebroadcast purposes, their quality left much to be desired. The convenience and advance in production techniques that a video tape recorder would bring to broadcasting—and eventually to nonbroadcast use—made the development of a usable machine mandatory. Hence, considerable time and effort was put into this development.

The early attempts to record a tv picture on tape simply adapted audio methods, with a few modifications to cope with the much wider frequency range of the video signal. A few simple calculations showed that a much higher tape speed and a narrower head gap were required, and, after a few experiments, several other problems arose which indicated that new techniques were required to record pictures of broadcast quality.

In the late 1940s, Ampex and RCA in the United States, and the BBC and Decca in Britain, did much work on the development of vtr's. The first results were various longitudinal machines that looked like large audio recorders. Although they did produce quite acceptable pictures, these machines had certain inherent faults which seriously limited their use and convenience. The main disadvantages were the large reels of tape and the very high tape speed. Maintaining this speed required a complicated transport system that was difficult to manufacture and maintain. The machine was inconvenient to operate, and it used an excessive amount of tape. The nature of the machine led to limited frequency response and dynamic range, which, coupled with the other objections, made it less than an optimum device for broadcast purposes.

In early 1955, the BBC actually went on the air with their VERA (Vision Electronic Recording Apparatus). It had a head gap of 20 microns and a tape speed of 200 inches per second. The tape for a half-hour program required a reel about 5 feet in diameter. The machine was large and expensive, and its quality was good but not perfect.

The limitations of these early machines were considered severe enough that some other method would have to be devised, or the vtr was doomed to be a device used only for very short programs or for inserts. In the early 1950s, work began on a revolutionary idea, that of a rotating head.

The important thing in tape recording is not the tape transport speed, but the speed of the head relative to the tape, or the *scanning speed* as it is sometimes called. A high scanning speed can be achieved by moving the head across the tape from edge to edge, rather than in a direction parallel with the length of the tape. Although at first this seemed to be a tremendous complication introducing further difficulties, the idea did present one sensible alternative which could either overcome the objections or simply bypass them.

By having the head sweep across the tape from one side to the other, the head could be made to record one field on one pass, the next field on the next pass, and so on, thus producing a continuous recording. Since there would be difficulties in wrapping the tape completely around the head, a system of multiple heads in a spinning assembly was used. In this way, the successive fields were recorded by successive heads. (For several practical reasons, broadcast machines actually record only part of the field with each sweep of the head, and several sweeps are used to record a complete field on the tape.) This method allowed a reasonable longitudinal tape speed and a high writing speed, but it required a wider tape than that used in audio.

Although this approach brought reel size and tape speed into practical limits, it did present a few problems quite new to the sphere of tape recording. The heads must rotate at a constant and accurately controlled speed, and they must be switched in and out of the recording circuit at exactly the right time in the tv signal and at the correct place on the tape. On playback, the heads must be made to align exactly with the recorded tracks; otherwise, no output will occur. The contact between the heads and the tape had to undergo much improvement, as did the tape oxide coating, which caused heavy head wear and initially would not record the high frequencies required.

Gradually, all of these problems were solved, and Ampex was able to produce its quad-head recorder in 1956. The quad machine was quickly adopted by broadcasting organizations throughout the world and has remained their standard and major choice for video recording. Its quality is superior to that of the kinescope film and is, in fact, equal to that of the live transmitted picture. Over the years, the recorder has been improved, and machines of this type are now manufactured by RCA, Shiba, and Fernseh as well as Ampex. There is complete compatibility among all machines, even with color recordings, provided that the same tv system is used.

Since the quad-head machine was developed by one group for a particularly specialized professional market, it was easy to get some standardization. However, this type of vtr is far too large and complex for anything but broadcast use. It is out of the range of any but the broadcasting profession, being too costly and complicated for general industrial, educational, or home-entertainment use.

It was not long before these potential nonbroadcast users were presented with an interesting possibility. In the early 1960s, the helical, or slant-track, machine appeared. However, helical machines were developed by different groups in various places in the world and at different times, with the added complication that some of the groups have changed their minds at times. The result is that standardization has not occurred. Without a single industrial market to set a standard, and since there was no precedent in format, a variety of formats and possibilities appeared. Because some of the work was performed in secrecy and isolation, a free exchange of ideas has not really materialized, and because the market has been most competitive as well as diverse and large, the different formats and types have all been able to get some foothold. The result is that several different formats of helical machines are now in popular use.

Certain aspects of video recording are hard to change, and these emerge as nearly standard ways of effecting a video recording. Such items are the adoption of fm recording, the necessity for control pulses and servos, etc. Nevertheless, it is in the details that most diversity is found, and these details do cover all areas of vtr technology. Recently, it appears that the Japanese manufacturers have settled upon a format to which they will all adhere, but for many years, many types of helical vtr's will be in existence and in ever-increasing numbers.

11

2

Review of Audio Recording

The tape recording process relies basically on four principles. These are simple principles, but a few additional facts of nature must be recognized and dealt with before high-quality recordings can be made and played back.

THE PRINCIPLE OF THE TAPE RECORDER

A modern tape recording is just a long, thin magnet formed from an iron-oxide powder glued to one side of a plastic tape. The intensity of magnetization varies along the length of the magnet, or, in other words, the strength of the magnet varies along the length, as indicated in Fig. 2-1. This is the first principle on which the tape recorder relies.

The second principle is that if an electric current passes through a piece of wire, it causes a magnetic field to surround the wire. If the wire is wound in the form of a solenoid, as in Fig. 2-2, then the solenoid acts like an ordinary bar magnet. The strength of the magnet is proportional to the magnitude of the current in the solenoid wire. If the current is varied, then the strength of the magnet varies.

The third principle is that if a magnetic substance, like soft iron, is placed in a magnetic field, it will become magnetized. If a second piece of material is placed in the field and the strength of the field is changed, then the strength of the second magnet will be different from the strength of the first.

These three principles may be carried to a logical conclusion: If the current through a solenoid is varied, the strength of the effective

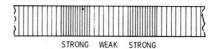

Fig. 2-1. Variation of magnetization along the length of a tape.

STRONG   WEAK   STRONG

magnet is varied, and so is the magnetic field around it. The residual magnetism of a ferrous material moved through the field also is varied.

In tape recording, a record head is the solenoid, the information we wish to record is the current in the wire, and the tape is the ferrous material which is moved past the head.

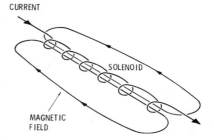

CURRENT

Fig. 2-2. Magnetic field around a solenoid.

SOLENOID

MAGNETIC FIELD

A fourth principle of interest is that the reverse of the preceding action is true. In other words, if a magnet is moved past a solenoid, a current will be produced in the wire. If the strength of the magnet is changed, the strength of the current will change. Thus, if a tape is pulled past a head, the output will be a changing current that represents the magnetic pattern on the tape.

RECORDING A SIGNAL ONTO TAPE

The tape is moved past the recording head at a constant speed, and it is kept in constant contact with the head, especially at the area of the head gap. A typical head is shown in Fig. 2-3. The head is shaped similar to a horseshoe magnet, with its flux concentrated in the gap. The record current from the electronics is fed to the coils in the head, and this current produces the magnetic field at the gap. The head is made of a magnetically soft substance so that it can be magnetized and demagnetized very easily and can respond to the recording current. The tape oxide is a hard magnetic substance that will retain indefinitely the magnetism imparted to it by the heads.

Ideally, the record gap should be as wide as possible (1-10 mils). This would allow as much magnetic flux as possible across the gap. The gaps are shaped to concentrate the magnetic flux as much as possible, and to allow it to distort its path through the tape oxide.

13

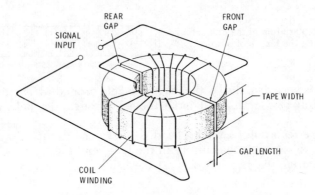

Fig. 2-3. Construction of a recording head.

The oxide provides an easier path for the flux than does the non-magnetic material used as a head-gap spacer.

The gap must be perfectly constructed at the edges, especially where the tape leaves the gap, because this determines the regularity of the recorded pattern on the tape. Any imperfections here should be less than 1/10th or even 1/100th of the shortest recorded wavelength.

The heads are made from many thin laminations (Fig. 2-4), wrapped with several hundred turns of wire. The changing current in the wire causes small magnets to be formed on the tape. The recorded magnetic flux pattern closely resembles the variations of the original current, and on playback the voltage output from the heads also resembles the current variations. Thus the reproduced output current is the same as the input current, except that it is much weaker.

In practice, the head gaps are approximately 0.75 mil. The coils have about 1500 to 2000 turns each, with one coil on each half of the head, giving an inductance around 150 mH. Ideally, the output

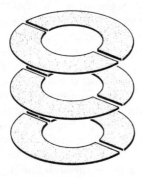

Fig. 2-4. Head laminations.

impedance of the head should be low to avoid losses and to keep the load constant across the frequency band.

HEADS

Heads are usually made from metals such as mumetal or permalloy. They are constructed from thin laminations (generally less than 4 mils thick) cemented into a stack; the coils are wrapped around this stack. The use of laminations reduces eddy-current losses in the metallic core of the head.

Recently, heads have been made of ferrite materials. A ferrite is a combination of ferric oxide with the oxides of other metals such as manganese, zinc, magnesium, and nickel. There are several properties of ferrites that make them ideal for recording heads. Magnetically they are soft, and hence they can change their magnetization easily with the recording current. They have a high sensitivity and low eddy-current losses, and so they do not need to be laminated. Their permeability is high, and they have low saturation densities.

Ferrite heads are not as efficient at low frequencies as conventional heads, and their acceptance in audio work has been limited. However, their high-frequency characteristics are excellent, so they have found much application in high-speed duplicators and particularly in video. As both of the latter uses involve a high tape-to-head speed, the wear factor is important, and in this respect ferrites are ideal.

Ferrites are very hard and brittle and must be treated with care. They are capable of taking a polished surface, but this can be damaged quickly by tape abrasion.

RECORDING BIAS

The early attempts at recording a signal on tape made use of the direct method of applying the desired signal to the head and moving the tape past the head at a constant speed. Although it is possible to record information in this manner, the results are unsatisfactory for several reasons. Only a very weak signal can be put on the tape before it becomes distorted, and it is accompanied by much noise and further distortion on playback. This distortion is due to the nonlinear characteristics of the tape.

When a magnetic force is applied to a magnetic material, the amount of magnetism acquired by the material can be determined from a graph such as the one in Fig. 2-5. As can be seen, this is a very nonlinear relationship, and there is a maximum limit of magnetization. The magnetizing force is assigned the symbol H, and the

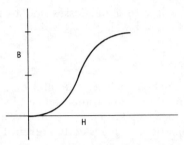

Fig. 2-5. A B-H curve.

magnetization of the material is assigned the symbol B. Hence, the curve is known as a *B-H curve*. When the force is removed, the material remains magnetized, but with less strength than when the force was present (Fig. 2-6).

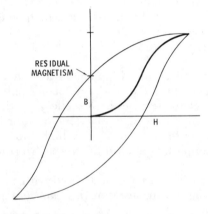

RESIDUAL
MAGNETISM

Fig. 2-6. A hysteresis loop.

If the force is now applied in the opposite direction, the magnetization can be reduced to zero, and the material can be magnetized in the opposite direction; i.e., the north and south poles of the magnet are interchanged. By reversing the force again, the original conditions can be repeated. This characteristic type of loop (Fig. 2-6) is known as a *hysteresis loop*. Is is important in tape recording because the magnetic force is the signal current in the heads, and the magnetic material is the tape.

It is due to this extremely nonlinear shape that the signal is distorted. Fig. 2-7 shows very small signal currents on the linear part of the curve. They are not distorted, but they are of such small amplitude that they are not much above the noise level. Fig. 2-8 shows the distortion in a larger signal. If the signal could be centered on the linear part of the curve, as in Fig. 2-9, it would be undistorted. Applying a constant magnetizing force along with the varying signal would accomplish this. A dc current through the

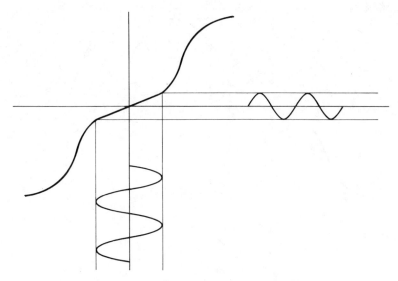

Fig. 2-7. Recording signal of small amplitude.

coils provides such a constant magnetizing force, and this was the first method used to obtain a higher level of undistorted recording on tape. The same method is used in many of the inexpensive audio cassette machines now available. Its main drawbacks are that it is noisy and the frequency response is limited to about 6 kHz at best.

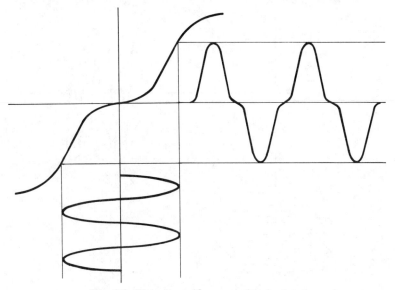

Fig. 2-8. Distortion of large-amplitude signal.

17

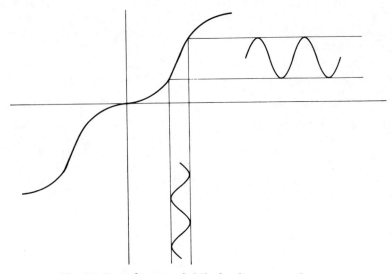

Fig. 2-9. Recording signal shifted to linear part of curve.

Fig. 2-10 shows how an ac signal in the form of a pure sine wave can also be used. The amplitude must be such that the peaks reach into the linear region. If the bias current is now varied by the signal current, it is possible to make a high-quality recording with low

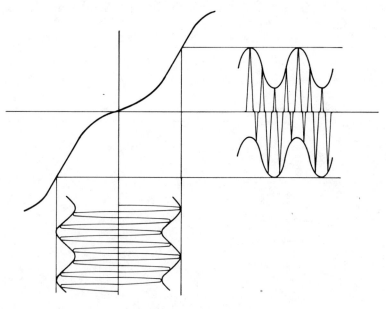

Fig. 2-10. Use of ac bias.

noise, low distortion, and extensive frequency response. This is the method adopted universally for use in broadcast and other high-quality machines.

The ac bias must be high enough in frequency that the harmonics of the signal will not beat with the bias to produce difference frequencies in the audio range. Since the second and third harmonics are the strongest, this indicates that the bias frequency should be above 60 kHz in a high-quality machine. Professional machines typically use 75 kHz; a home unit might use a frequency of about 40 kHz. A 5-to-1 ratio of bias to signal is considered optimum.

The bias signal is also used to erase the tape, and so the frequency must not be so high as to affect the erase characteristics adversely. The same oscillator must be used; if another is used, the two frequencies could cause audible beats on the tape.

The exact manner in which bias works is open to doubt. Several theories exist, and they will not be covered here. However, experience dictates that a clean sine wave must be used; otherwise, distortion and noise will result. The minimum level of bias current must be about 3.5 times the signal current, and it must be set to an optimum level for best results.

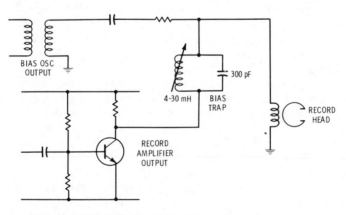

Fig. 2-11. Method of applying bias and signal to head.

The bias is generated in an oscillator and fed to the record head directly, where it mixes with the signal current (Fig. 2-11). The bias trap is a tuned filter to prevent the bias signal from getting into the record amplifier.

On playback, the bias signal is too high in frequency to be resolved by the reproduce head, and so it does not appear at the output of the head.

PLAYBACK

In playback, the tape passes the heads, and the magnetization on the tape causes a current in the head coils. This produces a voltage at the output terminals that is given by

$$v = kNR$$

where,
   k is a constant,
   N is the number of turns of wire in the coils,
   R is the rate of change of flux.

This expression is often written in the alternative form

$$v = N \frac{d\phi}{dt}$$

where $\frac{d\phi}{dt}$ is the rate of change of flux.

What this means is that the output of the head is directly proportional to the rate at which the magnetic flux is changing. If the frequency recorded on the tape is doubled, then the rate at which the flux changes is also doubled. Thus the output voltage from the head is doubled. A doubling in frequency is an increase of one octave, and a doubling in voltage is a rise of 6 dB. Hence, the output of a head rises typically at 6 dB per octave. Thus, in the playback of a tape, equalization is needed.

Another reason why equalization is needed is that the maximum output from the head occurs when the recorded wavelength is twice the head gap length, and zero output occurs when the wavelength is equal to the gap length. So the 6 dB per octave rise continues until a certain point is reached, and then in the next octave the output suddenly drops to zero (Fig. 2-12).

Beyond this point, a series of maxima and minima occur. Ideally, these are all of the same height; in actual practice, the peaks drop at about 4 dB per octave. The extinction frequencies all occur when the gap length is some whole-number multiple of the wavelength.

As an example, consider a tape speed of 15 inches per second (in/s) and a 20-kHz signal. The velocity, wavelength, and frequency are related by the expression

$$v = \lambda f$$

where,
   v is the tape speed (velocity),
   λ is the wavelength of the signal on the tape,
   f is the frequency of the signal on the tape.

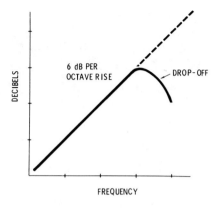

Fig. 2-12. Playback frequency characteristic.

Thus

$$\lambda = \frac{v}{f}$$

$$= \frac{15}{20 \times 10^3} = 0.75 \text{ mil}$$

This means that the wavelength of a 20-kHz signal on a tape traveling at 15 in/s is 0.75 mil. A head gap of this size will make 10 kHz the maximum-playback frequency and will make 20 kHz an extinction frequency. Thus in audio, gaps of 0.5-1.0 mil are quite common.

Some other factors now become important. Although a small gap is required for high-frequency response on playback, the smaller the gap, the less flux is available across this gap for recording. This is important in a head used for both purposes.

The high frequencies are also limited by the resonant frequency of the head. This frequency is determined by the inductance and stray capacitance of the windings.

Ideally, the head gap on playback should be much smaller than that used for recording; 0.0001 inch is ideal. Therefore, in a combined head a compromise is reached. A playback output of about 1 millivolt is typical, and the head should work into a high-impedance preamplifier. A further complication on playback is that the losses inherent in the head become significant.

LOSSES

The overall recording and playback system suffers from losses of signal due to several factors. The most apparent of these are the resistance of the coil windings, eddy currents in the material of the heads, and the various capacitance losses which tend to vary with

21

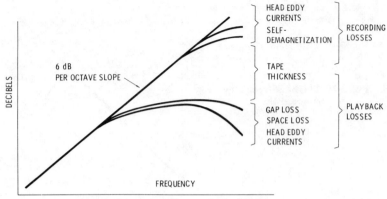

Fig. 2-13. Recording and playback losses.

frequency and construction. Fig. 2-13 shows the effects of recording and playback losses.

A head core made from a single thick lamination can have 5 to 10 dB of signal loss at 15 kHz, whereas a head made of many thin, well-annealed mumetal laminations or of ferrite will have lost only 1 to 5 dB at 1 MHz or higher.

A coil with a few turns of wire will have minimal RC losses, but will also have a low playback output. So, for a high playback output, many turns of wire are used. But such a coil can have so much stray capacitance that it will resonate with its own inductance, and the resonant frequency can be below the highest signal frequencies of the system; this can cause high losses.

Most losses occur at the high end of the frequency spectrum, and their causes can be classified as follows:

A. Finite gap of playback head
B. Demagnetization of the tape
C. Separation of the head and tape
D. Head alignment (i.e., out of azimuth)
E. Coating thickness of the tape
F. Field pattern of the record head
G. Eddy-current losses in the playback head
H. Eddy-current losses and hysteresis losses in the record head

Low-frequency losses are due almost entirely to the low rate of change of the magnetic flux past the head gap.

EQUALIZATION

Because of the characteristic rising output of the playback head, and the various losses suffered by the overall system, a recording in-

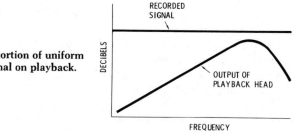

Fig. 2-14. Distortion of uniform recorded signal on playback.

put signal would suffer severe frequency distortion of playback. Fig. 2-14 shows this.

To overcome these losses and make the playback signal exactly like the original, the electronics must have a frequency characteristic that is the mirror image of these losses. Therefore, in general the frequency response of a tape playback amplifier is like that shown in

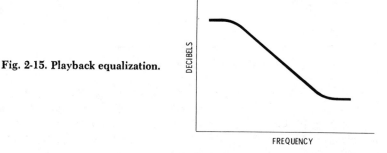

Fig. 2-15. Playback equalization.

Fig. 2-15. For different tape speeds, this curve has to be reshaped, because the slower the speed the greater are the head-gap lossses on playback. Some of the equalization can be accomplished on record, but it is mostly confined to the playback section.

Equalization of the electronics is most usually accomplished by the use of RC networks and feedback paths. These can be designed to give almost any shape required. Several standard shapes of play-back curve exist, the most common being the NAB, the CCIR, and the Philips Cassette standards.

AUDIO TAPE RECORDERS

When tape machines were adopted by the broadcast industry, it was relatively easy to set standards of performance and agree on items like tape speed, tape width, etc., so that an easy interchange of recordings was assured. The choice of equalization curves was not so simple, and several exist throughout the world. However, this does not prevent a tape recorded with one equalization characteristic from being played on a machine with another. As recorders began to spread into home use, most of these items of contention had been settled, and it was thus easy to make simplified copies of existing professional machines for domestic and other uses.

The head layout of an audio machine is almost a standard, since little variation from the basic layout of Fig. 2-16 is possible. Some machines have been built with heads facing away from the front of the deck. Professional machines have speeds of 7½ and 15 in/s (older models go up to 30 in/s). Smaller machines generally have speeds ranging from 7½ down to 1⅞ in/s. As the speed is lowered, the high-frequency response drops, and thus the equalization must be changed with the speed. Although a studio recorder presents no difficulty in recording and playing back 20 kHz, a small domestic cassette machine has an upper response limit of about 7.5 kHz.

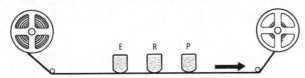

Fig. 2-16. Head arrangement for a three-head machine.

The electronic circuits have to provide two main functions, those of recording and playing back the signal. Professional machines have a record amplifier, a bias oscillator, and a playback amplifier. The nonbroadcast or domestic machine has an amplifier which is used in both the record and playback functions.

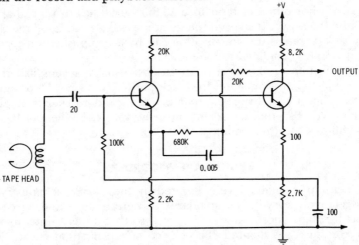

Fig. 2-17. Transistor tape-recorder amplifier.

A typical output from a microphone is around 1 millivolt, which must be raised to a level suitable to modulate the recording bias and to be put onto tape. This is fairly easy to achieve. All that is required is an amplifier with sufficient gain and a flat frequency response. This can be realized with three or four transistor stages, or even one IC chip.

In playback, the same or a similar amplifier can be used (Fig. 2-17). The output from a tape head is typically 1 millivolt, and this must be amplified to about zero level (0 dBm) in a preamplifier, which is also used to shape the frequency response. From the output of this preamplifier, the signal can be fed to a distribution line, as in broadcasting, or it can be used as the input to a speaker amplifier. A speaker amplifier is not always provided as an integral part of a tape recorder.

CONCLUSION

The audio range of 20 Hz to 20 kHz is about 10 octaves, and although the first machines could not handle this range, later machines could. However, this is about the limit that can be handled satisfactorily by this recording technique.

Over the years, magnetic tape has appeared in a variety of formats with many track layouts and speeds, and with many specialized machines of limited applications. Many different machines have been designed and used, from the massive Marconi-Stille with its 4-foot reels to the miniature cassette recorder of today with its IC electronics. Today, there is no doubt of the usefulness of the tape recorder to society.

From the original idea of recording sound, the tape recorder has successfully been used in applications that are dramatically different. These include data storage and processing, computer programming, and numerical control of machines. All of these applications were developed after the audio recorder had proven and established itself. The audio recorder did, in fact, provide a solid test bed for these new ideas. The video tape recorder grew out of the audio machine, and perhaps the most spectacular demonstration of how far it has progressed is the Mariner pictures from Mars. These were recorded on tape and then later played at a different speed for transmission to Earth, where they were computer processed to provide excellent pictures of the surface of the red planet.

3

Principles of
Video Recording

In principle, the video tape recorder is essentially the same as the audio recorder, utilizing magnetic means to record a signal onto a tape. However, the audio and video signals themselves are quite different. The most obvious difference is the frequency range. Audio extends from about 20 Hz to 20 kHz, which is a span of about 10 octaves. Video ranges from about 30 Hz to 4.5 MHz, which is around 18 octaves, and this is an almost impossible bandwidth for a tape machine to handle using the same means employed to record audio.

FREQUENCY RANGE LIMITATIONS

The gap effect is the most serious single restriction on the high-frequency response of a recording, and is of great importance in video recording. Consider a constant-current signal that spans the full required frequency range. If this signal is recorded onto a tape and then played back, the output of the head will increase at a steady rate as the frequency increases. This rise in output will continue until a maximum is reached, when the output will suddenly drop off to zero. This sudden drop is characteristic of all tape machines, and it limits the range they can handle to about 10 octaves. Beyond 10 octaves, the high and the low frequencies require too much equalization in the record and playback process. The head gap for audio is about the minimum that can be narrow enough for the high frequencies and still not have too much loss at the low end of the spectrum. Fig. 3-1 is a graphic explanation of this.

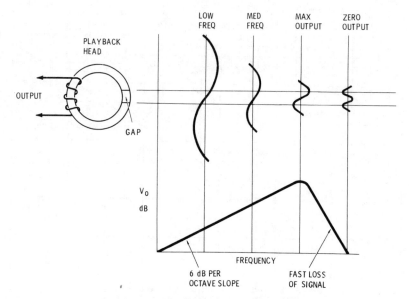

Fig. 3-1. Effect of head gap on output level.

A head designed for maximum output at 20 kHz has an extinction frequency to 40 kHz. Since the response goes down through the 10 octaves at a 6 dB per octave slope, the output at 20 Hz will be about 60 dB down. Compensation for this very wide spread in level is difficult, though not impossible.

The video signal ranges from 30 Hz up to 4.5 MHz. A head designed for maximum output at 4.5 MHz will have its output about 110 dB down at 30 Hz. This is far too wide a range to be handled by the same means as in the audio recorder. If the 4.5-MHz portion of the signal is not to overload the tape, then the 30-Hz to 15-kHz range is far down in the noise. If these lower frequencies are recorded at a reasonable level to ensure a good signal-to-noise ratio, then the tape will overload at about 1 MHz. As the high frequencies represent picture quality and the low end is required for the sync pulses, both are equally important to a good recording. These frequency and dynamic ranges are illustrated in Fig. 3-2.

The 10-octave range cannot be extended or improved by varying the tape speed or the head gap. Increasing the speed and reducing the head gap improves the high-frequency response and allows the maximum frequency to be increased, but it impairs the low-frequency playback response.

Consider a typical audio situation with a top frequency of 20 kHz, a tape speed of 15 in/s, and a single record/playback head.

27

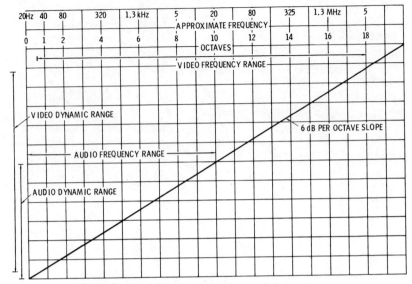

Fig. 3-2. Audio and video frequency and dynamic ranges.

$$\lambda = \frac{v}{f}$$

where,
    v is the tape speed,
    f is the recorded frequency,
    λ is the recorded wavelength.

In this example,

$$\lambda = \frac{15}{20 \times 10^3} = 0.75 \text{ mil}$$

This means that the wavelength of a 20-kHz signal on a tape traveling at 15 in/s is 0.75 mil. If the head gap is 0.75 mil, then 20 kHz is an extinction frequency, and the maximum output on playback will occur at 10 kHz. Thus in audio, head gaps of 0.3 mil and less are common. A good high-frequency response dictates a small head gap, but if the gap is too small, then insufficient signal is recorded on the tape, and the low-frequency level is down in the noise on playback.

Now consider a 5-MHz video signal with a tape speed of 15 in/s.

$$\lambda = \frac{15}{5 \times 10^6} = 3 \text{ microinches}$$

Such a small gap is achievable but is not really practical. Although

28

it would handle the high frequencies, the lows would be changing too slowly as they crossed the gap on playback, and thus their output would be very low.

Now consider a 5-MHz signal and a tape speed of 150 in/s. This gives a wavelength of 30 microinches, so increasing the speed will obviously allow a wider head gap to be used for the same high-frequency response, and this wider gap is better able to handle the low-frequency response on playback.

However, the problem cannot be completely solved by increasing the tape speed, as this does not overcome the 6 dB per octave loss on playback. For every octave of high frequency gained, an octave of low frequency is lost into the noise region. Thus if the gap is made long enough to accommodate the low frequencies, it will be too wide for the highs on playback, and if it is made narrow enough to give a good high-frequency playback response, then the low frequencies are too low in level. Across the audio range, a good compromise in gap width is possible, but the video bandwidth is far too wide to be accommodated by this technique.

To summarize: The maximum frequency that can be played back is determined by the head gap and the tape speed, and the bandwidth that can be reproduced is limited to approximately 10 octaves by the nature of the reproducing system. Hence, for an 18-octave spread of video signal to be recorded and reproduced, some other technique is required.

THE METHOD ADOPTED

The method most easily adopted is to translate the 4.5-MHz spread up the frequency spectrum so that although the spread still remains the same, the ratio of the highest frequency to the lowest frequency is much reduced. Having the frequency swing from 5 to 10 MHz still represents a 5-MHz bandwidth, but the range is now only one octave instead of 18. A head gap designed for a 7.5-MHz optimum would handle the 5-MHz and 10-MHz frequencies equally well and would require very little equalization. Once a frequency translation of this form is effected, the only way left to use it is to have a signal which varies in frequency between two extremes. This is, of course, a frequency modulated (fm) signal.

An fm signal can be impressed onto a tape in the same way that an unmodulated bias is applied in an audio recorder. It can be made strong enough to saturate the tape, thus ensuring a good signal strength. Since the demodulation of an fm signal removes all amplitude variations, the recording is made impervious to much noise. In this fashion, the 6-dB difference between the 5 MHz and 10 MHz points shown in Fig. 3-3 can be completely eliminated, thus removing

29

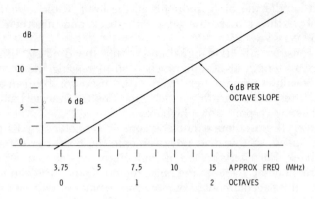

Fig. 3-3. Level variation for fm recording.

the need for any playback equalization. This method avoids the non-linearities introduced by the tape transfer characteristics.

In practice, a high-frequency sine wave of sufficient amplitude to saturate the tape is used. The frequency of this signal is varied by the video signal, with the lowest carrier frequency corresponding to the sync tips, and the highest carrier frequency corresponding to peak white (Fig. 3-4). The actual frequencies chosen are not critical, and they vary from machine to machine. In most helical machines, the fm carrier is varied from about 2 MHz to 5 MHz, and sometimes it ranges as high as 10 MHz. The total fm spread is never more than about 4 MHz. Some of the later helical recorders use a square wave instead of a sine wave. The square wave is produced by a multivibrator, which is easier to frequency modulate than a sine-wave oscillator.

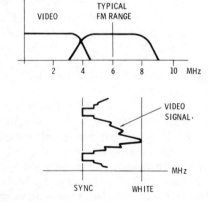

Fig. 3-4. Frequency deviation in video recording.

Another good reason for adopting an fm carrier is that the amplitude fluctuations caused by erratic contact between the heads and the tape do not affect the signal nearly as much as they would if an am signal were used. Thus the dropout problem is minimized. Using double-sideband fm with a large modulation produces less distortion on playback than would using am, and it also keeps the modulation noise very low, which further improves the signal-to-noise ratio and improves the quality of the picture.

The use of an fm carrier solves the electronic equalization problems associated with a video signal by reducing its bandwidth from 18 octaves to about 2 octaves. However, since a carrier frequency in the low megahertz range is required, a high tape speed and small head gap are still necessary. If the tape speed is increased while using the same transport system as an audio tape recorder, a large quantity of tape is used in a very short time, and several other difficulties still exist.

Consider a practically achievable gap of 0.1 mil and a carrier frequency of 5 MHz. After this highest playback frequency is decided and the minimum practical head gap is established, the necessary tape speed (v) can be calculated:

$$v = 2GF$$

where,
G is the head gap,
F is the highest playback frequency.

In this case,

$$v = 2 \, (0.1 \times 10^{-3})(5 \times 10^{6})$$
$$= 1000 \text{ in/s}$$

Thus, even with a high-frequency fm signal and a reasonable head gap, a high tape speed is still required. With this tape speed the number of feet of tape (L) used per hour can be determined:

$$L = \frac{1000 \times 60 \times 60}{12}$$
$$= 300,000 \text{ ft/hr}$$

This is an excessive amount of tape. For a 1-mil tape thickness and a 6-inch inner diameter, the reel required would have a diameter of about 4 feet. Such a reel is too large and cumbersome for comfortable operational handling; the amount of tape used makes this very expensive; and the mechanics required to transport the tape safely and accurately are complex, costly, and unreliable.

Consider the following two situations:

A. A 2-MHz signal bandwidth and a 0.1-mil head gap.

$$v = 2(0.1 \times 10^{-3})(2 \times 10^6)$$
$$= 400 \text{ in/s}$$

This is still a very high speed.

B. Consider now using a speed of 200 in/s, which is about the maximum achievable with ease, and the same head gap. Since

$$v = 2GF$$

then,

$$F = \frac{v}{2G}$$

In this case,

$$F = \frac{200}{2(0.1 \times 10^{-3})}$$
$$= 1 \text{ MHz}$$

This is a carrier frequency of only 1 MHz, and this is too low a frequency to be modulated the full range required for a video signal. This would produce pictures totally inadequate for a broadcast organization, and they would be seriously lacking in quality even for a low-grade cctv system.

From this, it is apparent that an audio-type longitudinal machine would have several problems and would be lacking in quality. To overcome these mechanical problems of a small head gap, high tape speed, and excessive size of reels, a method of splitting the video signal into sections was tried. This solved the problem of an excessive bandwidth to be recorded, since each band could now be chosen for near optimum performance and be separately modulated. Some of the early experimental machines used this technique with considerable success.

In 1955, the BBC in London actually went on the air with such a machine, named VERA, developed by the BBC and Decca. It had a relatively short and semisuccessful life until it was eclipsed by the advent of the quad-head machine. It utilized the longitudinal method of recording, splitting the video band into two sections and applying them separately to the tape.

Band A contained the sync and video information up to 100 kHz. This portion of the video signal was made to frequency modulate a 1-MHz carrier which was made free from all amplitude variations. Band B was for the video information between 100 kHz and 3 MHz.

(The British 405-line system requires only a 3-MHz bandwidth.) This signal was recorded using an am carrier. Band C carried the audio information, which was made to frequency modulate a 250-kHz carrier.

The three identical heads were made of ferrite, each with the same gap of 2×10^{-5} inches. The record heads were encapsulated in one unit with copper screens between them for shielding. The playback heads were identical to the record heads, except they had different windings. This assembly was mounted a few inches from the record-head assembly and gave the advantage of the capability for immediate playback.

The tape used was ½ inch wide and was mounted on reels 20½ inches in diameter. This produced only 15 minutes of recording time. The tape speed was 200 in/s, and the spools were motor driven at just below this speed so that little power was needed by the capstan to maintain an exact 200 in/s. The capstan motor was a split-field dc motor servo controlled to run at 3000 rpm. The servo was referenced to the power frequency of 50 Hz, and feedback from the capstan was provided by a phototach system.

Although VERA was a physically large machine, it was not enormous beyond reason. The reels could be handled fairly easily by an operator, and they were mounted vertically on the front panel.

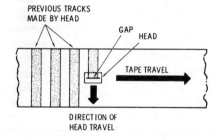

Fig. 3-5. Transverse recording by movement of head across tape.

In general, the longitudinal machines had several problems that could not be overcome. The high tape speed was very hard to control accurately, because it required a very complicated drive mechanism that was subject to much instability and other troubles. The machines all suffered from restricted bandwidth, used too much tape, and would not record for long periods of time. In nonbroadcast work, economy was a consideration, and much was pared from the necessary electronics and mechanics, resulting in general unsatisfactory performance. In later years, a few small longitudinal recorders appeared, but just as quickly disappeared.

Obviously, this was not a suitable method of video recording, and some other means had to be found. The answer to these problems was found in realizing that methods of recording on tape other than in the longitudinal direction existed and had in fact been used. One idea was to have the head move across the tape, as in Fig. 3-5. Having a head move in this fashion could produce a very high head-to-tape speed and still allow a reasonable tape-transport speed. The development of this idea produced the rotating-head recorder.

The advent of this machine produced a series of problems quite new to the magnetic-recording industry. A rotating head traced out a track across the tape. When it reached the bottom edge of the tape, it had to complete a retrace path before it could be used to record on the tape again. During this retrace period, another head was used to record the next track across the tape. Obviously, it was necessary to switch from one head to the next, so the concept of continual head switching was introduced.

On playback, the heads had to follow the recorded paths exactly as they were put on the tape, and it was necessary to switch from one head to the next at the correct time. This meant the heads had to be placed in the correct position with respect to the tape linear motion, making a servomechanism necessary to control them.

With a low linear tape speed, the tape did not travel far in the time it took one head to sweep across the tape width; therefore, the recorded tracks were close together and very thin. Due to this thin width of track, tape imperfections became important. This made the concept of space loss a practical nuisance instead of a theoretical idea, and thus the idea of tip penetration found its way into recording technology.

The nearly vertical tracks require a very steady tape tension and speed, particularly on playback. Because of this, the tape transport mechanism of recorders underwent a radical change and improvement.

Since the foregoing points are fundamental to video recorders, they are covered in more detail in the next few paragraphs.

SERVOS

The rotating heads are mounted on a motor-driven headwheel. On the same shaft is a sensing device that produces a series of pulses to indicate the position and speed of the rotating heads. In the record mode, the pulses from the sensor are recorded on one edge of the tape. In this mode, the tape is driven at a constant speed by a motor so that these pulses will be evenly spaced along the control track when the machine is running correctly. On playback, the control pulses are compared to the head pulses to position the head correctly in relation to the tape. This now enables the heads to follow the recorded tracks correctly and reproduce the video signal.

HEAD SWITCHING

As more than one head is required to record the full tv signal on a tape, a system of switching from one head to the next is needed. The switching circuitry is controlled by the pulses generated in the headwheel assembly so that switching occurs at the correct time and thus produces a continuous recording.

TAPE TENSION

On playback, the tension in the tape must be the same as it was in the record mode. If it is not, the spacing between the tracks will vary slightly, and the heads will not scan the recording properly. Various methods are used to control the tension, and these are covered in a later chapter. Since atmospheric conditions as well as minor machine problems can cause changes in tape tension, it is a subject of prime importance.

SPACE LOSS

In early audio recorders, the full width of the tape was used to record the signal. In later models, the width of the track was reduced so that two, four, and now even eight tracks are found on a ¼-inch tape. The narrower the tracks, the less signal is available for playback, and so the signal-to-noise ratio is impaired as the track width is lessened.

In each of these cases, there is enough tape area passing over the head gap so that a slight unevenness of the tape or a small hole in the coating does not seriously affect the signal. As the size of such irregularities is often small, the signal produced by them is of a very high frequency and at a level low enough to be masked by the recorded signal. Thus, minor tape imperfections are no real problem in audio recording.

In video recording, this is not the case. Small irregularities in the tape are often comparable in size to the wavelength recorded on the tape, so a small lump or hole can cause the tape to leave the head and present no signal to the head. This condition is made worse by the very thin width of the track, and the problem of uneven tape becomes serious. A loss of signal is usually observed as horizontal flashes on the screen, lasting from a fraction of a line to several lines and spread over several frames, since the imperfection is both wider than the track and at a sharp angle to it (Fig. 3-6).

This loss of signal is called a *dropout*, and its magnitude can be calculated as follows:

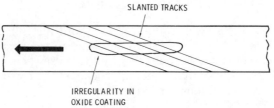

Fig. 3-6. Tape fault across slanted tracks.

$$\text{Signal loss in dB} = 55\frac{D}{\lambda}$$

where,
D is the distance of separation of the tape from the head,
λ is the wavelength of the signal on the tape.

In an audio recording, a typical situation might be a minor separation of 0.1 mil, caused by a piece of dirt or loose oxide, during the playback of a 7.5-kHz signal on a 7½-in/s tape.

$$\lambda = \frac{v}{f}$$

$$= \frac{7.5}{7.5 \times 10^3} = 10^{-3}\,\text{inch}$$

Thus

$$\text{Loss} = 55\frac{10^{-4}}{10^{-3}} = 5.5\,\text{dB}$$

This is not a serious loss and is within the bounds of practical possibility due to tape curl, rough surfaces, etc. But now look at the same imperfection in a video recording with a head-to-tape speed of 1000 in/s and a 5-MHz signal.

$$\lambda = \frac{1000}{5 \times 10^6} = 2 \times 10^{-4}\,\text{inch}$$

$$\text{Loss} = 55\frac{10^{-4}}{2 \times 10^{-4}} = 27.5\,\text{dB}$$

This is a large loss of signal and would be quite noticeable. It is outside the range of signal variation normally compensated for by the fm processing, and it is in fact just below the level at which automatic dropout compensators insert a substitute signal.

The graph in Fig. 3-7B shows this loss in decibels plotted against the ratio D/λ. As the curve shows, the losses incurred by signals of short wavelengths are serious.

These losses are caused by surface lumps on the tape, dirt on the heads, and irregular tape motion. An improper finish on the head near the gap can cause a loss of 10 dB at 15 kHz. Spacing loss can be as serious as azimuth loss at high frequencies. As the wavelength decreases, the excursions of the lines of flux from the tape surface are decreased, and so this increases the spacing loss at high frequencies. Therefore, in video recording, it is essential to maintain a perfect contact at all times. This contact is ensured by making the tip of the recording head actually penetrate slightly into the tape.

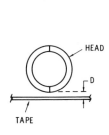

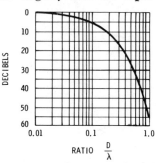

(A) *Separation of head from tape.* (B) *Graph of loss in decibels.*

Fig. 3-7. Relation of space loss to head-tape separation.

Since the head penetrates the tape slightly, it does in fact gouge out a thin channel as it scans the tracks. If the head penetrates too far or the tape tension is too great, the head will gradually wear away particles of oxide from the tape coating. These particles collect as a fine dust in the head mechanism and clog the head and hold it away from the tape, resulting in loss of signal. Some of the earlier tapes that had a loose coating caused this problem quite frequently, but later tapes have been improved to the point where they no longer do so. Generally, this is a small fault that can be remedied quickly by cleaning the head regularly and checking the tip penetration and tape tension. In practice, it is found that a head penetration of 40 to 100 microinches is about the optimum to guard against dropout and yet not cause other problems.

CONCLUSION

At first, it would seem that a machine of the type described in the last several sections was far too complicated to work reliably. The difficulties presented seemed almost insurmountable. However, due to the pressing need for a suitable machine, research and work continued. The result was the appearance of the quad-head recorder in 1956. At the National Association of Broadcasters convention in

Chicago, Ampex demonstrated their brand new Model 1000. It was an instant success.

The next chapter is a brief description of the modern quad-head recording machine. This is followed in subsequent chapters by a detailed examination of the helical vtr.

4

The Broadcast
Quad-Head Recorder

The considerations of the last chapter led to the development of the first practical video tape recorder. This was the Ampex 1000, which was the first to use a rotating head. It was designed as a broadcast machine, and its performance was aimed at the very tight specifications of that industry. Fig. 4-1 shows a modern broadcast quad-head machine typical of the many now in use throughout the world.

The machine uses a wheel (Fig. 4-2) with four heads mounted in it; the wheel rotates so that the heads scan tracks across the tape. The tape is arranged to curve slightly so that good contact is maintained at all times, and the physical dimensions are such that as one head is about to leave the tape another makes contact at the opposite edge. In this way, a slight overlap of video information is recorded, thus allowing a smooth playback of the complete picture. The recorder uses a two-inch-wide tape which travels at a speed of 15 inches per second. In this way, a full hour of program can be put onto one reel about ten inches in diameter.

THE HEAD ASSEMBLY

The head assembly refers to the rotating heads, their mounting, and the other facilities that are built into the same structure. The main part is an electric motor that drives the wheel in which the heads are mounted. They are mounted in the periphery of the wheel and rotate so that they travel across the width of the tape in a

Courtesy Ampex Corp.

Fig. 4-1. A broadcast-type quad-head video tape recorder.

direction at right angles to the direction of tape travel (Fig. 4-3).

The tape is held in a curved shoe, or guide (Fig. 4-4), by air suction. The purpose of the shoe is to ensure that the tape does not leave the heads and that correct tip penetration is maintained. The shoe is in effect a female guide that can be adjusted so that the centers of the tape and wheel are coincident. To ensure good contact with the tape and to overcome the effects of head wear, a small groove is made in the guide, and the head extends into it. The tape is stretched around the heads, and this also helps to reduce certain timing and other error effects. Guides are mounted on the edge of the shoe to set the tape so that the heads come into contact about 90 mils down from the upper edge.

If each of the heads recorded a full tv field as it scanned the tape, the time taken to scan the two inches would be 1/60 second. This would give a scan speed of 120 in/s, which is far too slow, because to get a sufficient bandwidth on this short track the head would have to be extremely narrow.

Courtesy Ampex Corp.

Fig. 4-2. A head assembly for a quad-head machine.

Since it is almost a physical impossibility for the tape to completely encircle the headwheel, some other arrangement is needed. This is why the four heads are used to scan the tape successively, and the tape is only slightly curved to fit the wheel.

An fm signal in the low megahertz region is put onto the tape. If a signal of approximately 7.5 MHz is assumed, with a head gap of 100 microinches, the head-to-tape speed can be calculated.

$$v = 2GF$$
$$= 2(100 \times 10^{-6}) \ (7.5 \times 10^{6})$$
$$= 1500 \ \text{in/s}$$

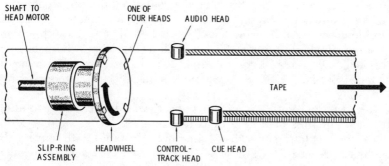

Fig. 4-3. Heads in a quad-head video tape recorder.

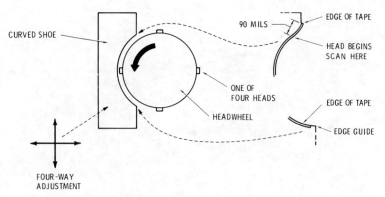

(A) *Relationship of shoe and headwheel.*

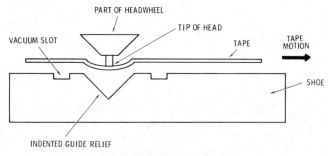

(B) *Cross section of shoe.*

Fig. 4-4. Curved shoe and rotating headwheel.

where,
    v is the head-to-tape speed,
    G is the head gap,
    F is the highest carrier frequency.

This is a very high head-to-tape speed, and to achieve it with the four heads, the headwheel is rotated at 240 revolutions per second.

The headwheel is approximately 2 inches in diameter; hence, its circumference is approximately 6.28 inches. If it is made to rotate at some simple multiple of 60 Hz, such as four times the field rate, this will give the 240 rps rotation. This works out to be a head-to-tape speed of 1507 in/s, which is sufficient for a good quality recording.

Because the rotation is at four times the field rate, one revolution puts one quarter of a field onto the tape. Since one field is 262.5 lines, one revolution puts 65.625 lines onto the tape. With four heads, each head records 16.4 lines in each track. Thus 32 tracks are needed for a complete tv frame.

In practice, the heads and tape make contact for between 113° and 120°, depending on the model of the machine. This means that an overlap of video information can occur, and this is used by allowing each head to record 18.6 lines of video in each track. This overlap is arranged so that the last two horizontal lines of each track contain the same information as the first two of the next track. During playback, head switching occurs during this overlap time.

A 360° rotation of the headwheel takes 1/240 second, so a 113° arc takes 0.001308 second. This is aproximately the time it takes one head to cross the width of the tape. At 15 in/s, the tape travels a distance, d, given by

$$d = 15 \times 0.001308$$
$$= 0.01962 \text{ in}$$
$$= 19.62 \text{ mils}$$

Thus the bottom of the track is displaced along the tape by this amount when compared to the top. This means the tracks slant at an angle of about 0.57°, or 33 minutes of arc.

In this method of recording, obviously the width of the track is dependent on the size of the scanning head. The narrower the track, the more picture information can be recorded on the tape. But too narrow a track will not produce enough signal on playback, and thus the signal-to-noise ratio will be impaired. The track width is chosen to be 10 mils, and the heads are constructed to this size. A guard band is placed between the tracks to give a definite separation. If this band is too wide, too much tape is used, and if it is too narrow, tracking becomes a problem. Tracking is the ability of the playback head to scan only one recorded track at a time and not to cross more than one. The effects of mistracking are very noticeable on the picture screen. A guard band of 5.6 mils is chosen, and the ratio of 2:1 for track to guard-band width is about optimum. Fig. 4-5 shows the full layout of the quad-head tracks, with the important dimensions indicated.

THE TAPE PATH

In appearance, the quad-head machine is not unlike a large audio recorder. The supply reel (Fig. 4-6) is placed on a hub on the left-hand side of the machine, and the tape passes from this reel over a full-width erase head. A guide associated with the erase head determines the angle at which the tape enters the shoe area; this is very critical in shape and a most important part of the tape deck. Once in the shoe, the tape is held against the heads while the video is re-

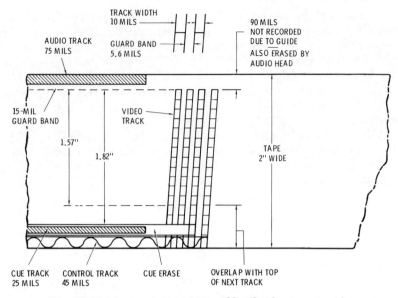

Fig. 4-5. Track arrangement for quad-head video tape recorder.

corded. The control-track head is placed ¾ inch after the center of the headwheel. Nine inches farther downstream are the combined audio/cue heads, which are preceded by their own erase heads. Finally, the capstan and roller feed the tape into the takeup spool. To keep the tape in its correct path, guideposts are used at strategic points.

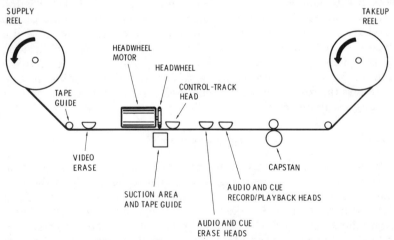

Fig. 4-6. Tape path in quad-head recorder.

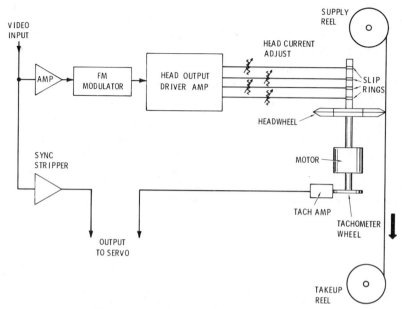

Fig. 4-7. Record mode of quad-head recorder.

RECORD MODE

Fig. 4-7 is a block diagram of the record process, showing only the record electronics and not the servomechanism. The input to the system is the composite video signal. This is first amplified, and then the higher frequencies are preemphasized to improve the signal-to-noise ratio. The gain of the amplifier is set to provide the desired controlled frequency swing in the fm modulator. The processed video signal is fed to the modulator with a preset dc level. The dc sets the unmodulated carrier frequency that is varied by the video signal. The modulated signal is fed to the heads simultaneously in parallel, with no electronic or mechanical switching used. Therefore, due to the physical arrangement of the heads and the tape, some duplication of signal occurs between adjacent tracks. Before reaching the heads, the signal passes through adjustable delay lines, one for each head, which can be set to correct for certain errors inherent in head manufacture. The outputs from the delay lines are applied to the heads through an arrangement of slip rings.

The fm frequencies used are typically from 1 MHz to 11.4 MHz in low-band monochrome recorders, and from 1 MHz to 13.6 MHz in high-band color machines.

Recording is a fairly straightforward process, with the heads and tape maintained at constant speeds so that the tracks are put onto the tape in an orderly and controlled manner. The tension in the tape is maintained constant in the record mode so that the control track is recorded evenly along the edge of the tape. This becomes important in the playback mode.

PLAYBACK MODE

A block diagram of the playback function is shown in Fig. 4-8. On playback, the signal from the heads is passed through delay lines and then into individual preamplifiers. The outputs of the preamplifiers feed the switching network, which automatically selects the scanned track and feeds it to the fm limiters. From here the signal is further processed and demodulated to a video signal, which is fed to the various outputs of the machine.

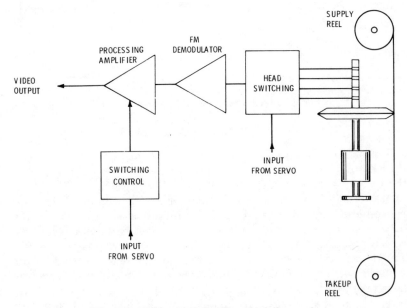

Fig. 4-8. Playback mode of quad-head recorder.

Playback is not as simple a procedure as record, as it is here that most of the problems which beset vtr's occur. Most of these problems stem from the fact that the heads must now accurately retrace the recorded video tracks. This means the tape has to be in an exact position at an exact time so that the heads can begin their scan across the tape and follow the path of the recorded track perfectly. Obviously, they must also scan across this track in exactly the same time it took to record the track.

An added complication is that if a signal is still being fed into the machine from some source for editing or insert purposes, the playback video must be synchronized with the incoming signal so that a switch from one to the other will not upset the tv screen.

Achieving perfect playback is not easy, and very complicated head and capstan servo systems are absolutely necessary. The record and playback electronics will not be covered here, because in principle they are the same as those in the helical machines, which are covered in a later chapter.

On playback, a tension servo is used to obtain the same tension as used on record. This servo uses the horizontal-sync pulses and the control-track pulses to control the back tension applied to the feed spool. A tension-controlling servo is necessary because factors such as the ambient temperature and humidity affect the tape considerably. However, the majority of the electronics and the mechanical structure is associated with the headwheel and its servo.

THE QUAD-HEAD SERVOS

Unlike the longitudinal recorder, the rotating-head machine cannot just record the incoming information randomly along the length of the tape; a definite pattern for laying down the tracks is now required. It does not matter how the video tracks are laid down, but on playback the heads obviously must trace the tracks exactly as they were recorded.

To facilitate this and ease the considerable problems involved, an exact place on the tape is chosen for the vertical interval, and definite places in the video signal are chosen for switching to the next track. Also, the linear speed of the tape and the rotation of the heads must be carefully controlled at all times. Since the tape must be in the correct position at the correct time for the heads to scan the tracks properly on playback, both the heads and capstan are servo controlled. Servos are described in more detail in Chapter 9.

HEAD SWITCHING

On playback, the heads are switched sequentially to give a continuous picture output. The switching point is placed on the front porch of a horizontal-sync pulse, since this is not seen on the screen. The selected output comes from one head at a time while the other three are turned off.

To tell the switch when to operate, a 240-Hz square wave derived from the head-tach pulse is used. This combines the outputs of heads 1 and 2 into one channel and heads 3 and 4 into another channel. Then a 480-Hz square wave combines these two into a final signal. This 480-Hz signal is timed by the horizontal information from the tape so that the switching is effected at the correct time. In this way a half cycle of the 480-Hz square wave corresponds to a 90° rotation of the headwheel.

After combination, the four signals are passed through a master equalizer to optimize the entire system.

QUADRATURE ERROR

In manufacture, it is almost impossible not to have some small errors in the placing of the four heads on the periphery of the headwheel. If the heads are not exactly at 90° to each other (Fig. 4-9), the effects of this *quadrature error* can be seen on the screen.

At 240 rps, 1° of rotation will take about 12 microseconds, and 0.01° will require 0.12 microsecond, or 120 nanoseconds. This is

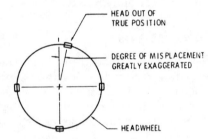

Fig. 4-9. Quadrature error.

equal to 36 seconds of arc, which is one picture element. If a horizontal line jumps sideways by one picture element, the effect can be clearly seen on the screen, so misplacement of a head by 36 seconds of arc will cause a horizontal jumping of the picture every time that head begins and ends its scan.

The heads can be adjusted so closely that this effect can be eliminated. A small screw is provided which will alter the head position so that the quadrature is correct. This is easy to do properly, since the results can be checked on the screen.

New heads are now set at the factory to be within 0.02 microsecond of time, or 6 seconds of arc. This is unnoticeable on the screen.

OTHER ERRORS

The four gaps of the recording heads must all be exactly at right angles to the direction of the tracks across the tape. A deviation from this orientation is called *azimuth error*. Wrong azimuth produces picture effects similar to those caused by quadrature error.

Skewing of the picture is caused by a difference in the head-tip penetration in the record and playback modes.

Scalloping is caused by differences in the height of the guides. It can be corrected by setting the guide heights with an alignment tape.

HEAD WEAR

As the machine is used, the head tips are gradually worn down by the abrasive action of the coating on the tape. New heads are set to penetrate into the tape by 3.7 mils, and a well used head may only penetrate by 1.0 mil. Reference to Fig. 4-10 will help in the following explanation.

With new heads, the total wheel diameter, D_n, is given by

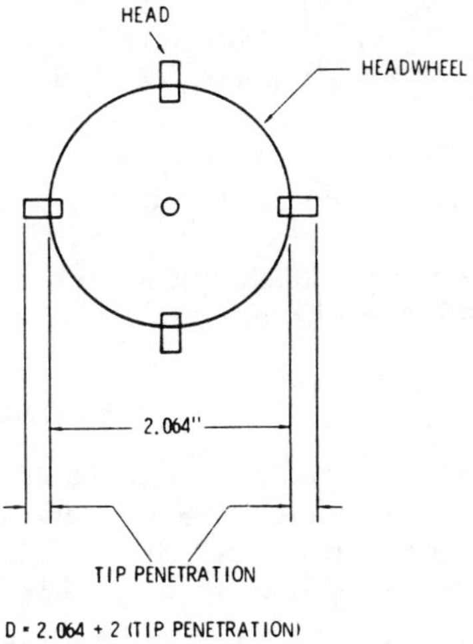

Fig. 4-10. Overall headwheel diameter.

$$D_n = 2.064 + 2(0.0037)$$
$$= 2.071 \text{ in}$$

Thus the circumference, C_n, is

$$C_n = 2.071\pi$$
$$= 6.50 \text{ in}$$

and the head-to-tape speed, V_n, is

$$V_n = 6.50 \times 240$$
$$= 1560 \text{ in/s}$$

With a worn head, the total wheel diameter, D_w, is

$$D_w = 2.064 + 2(0.001)$$
$$= 2.066 \text{ in}$$

Thus the circumference, C_w, is

$$C_w = 2.066\pi$$
$$= 6.49 \text{ in}$$

and the head-to-tape speed, V_w, is

$$V_w = 6.49 \times 240$$
$$= 1557 \text{ in/s}$$

This is not a great enough lessening of the speed to affect monochrome operation, and it is well within the capabilities of the servo to correct. But consider the situation for color. A difference of 3 in/s in a speed of 1560 in/s is 0.2 percent. The time of one horizontal line is 63.5 microseconds, and 0.2 percent of one line interval is therefore 0.12 microsecond. The peaks of the 3.58-MHz color reference signal are 0.28 microsecond apart. So a 0.2 percent variation in the tape speed represents almost half the wavelength of the 3.58-MHz reference. Half a wavelength is 180°, and as little as 5° of phase error can be detected by the eye. Therefore in color recording and playback, the servo must be able to compensate very accurately for all head wear.

CONCLUSION

The foregoing paragraphs are obviously a very brief description of the quad-head machine, and in no way do they cover everything which could be described. This machine is an incredibly complex piece of equipment that is not for untrained amateurs to use, and it most certainly requires a professional broadcast engineer to maintain and service it. It is as expensive as it is complicated, and it is not a machine to treat lightly or abuse.

To discuss fully the record, playback, and servo electronics as well as the mechanics would take too much time and space and would serve no purpose since this is not a commonly available market item outside of broadcast television. For the same reasons, the differences between models and manufacturers were not covered. Although this machine is excellent for color recording and playback and will perform perfect color editing, the manner of achieving and correcing color has also been intentionally omitted. This chapter was included to point out the important principles of vtr's that made the development of a simpler machine highly desirable.

5

The Helical VTR

The quad-head recorder does a superb job of video recording, but it is far too large and complex for anything outside of the broadcast industry. However, by 1960 nonbroadcast tv had developed to the point where a vtr had become essential, and for this type of tv work a much smaller, simpler, and cheaper machine was required. Such a machine made its first appearance in 1960. It relied on an idea that had been developed along with the quad machine, but had been discarded for broadcast use.

FUNDAMENTALS OF THE HELICAL VTR

Reviewing the requirements for this simpler machine indicated that the very critical heads and their mounting used in the quad-head machine would obviously need to be simplified, as would the very complicated electronics needed to support this type of mechanism. Due to the very high head-to-tape speed required, some form of rotating heads was still needed, and so also was some complex electronics.

In a quad-head machine, the transverse tracks on the tape are very short in length, and the heads take a short time to travel across the tape. Each single track contains about 16 lines of video, and four revolutions of the headwheel are required to record one field, which amounts to 16 passes of the heads (or 16 tracks) per field. A great simplification could be effected if only one pass of a head could record a whole tv field. Simply using the same format and slowing the heads so that this is accomplished does not work, because the head-to-tape speed becomes too slow to produce a good quality picture. Also, it is impossible to reduce the number of heads

Courtesy Sony Corp. of America
(A) *Sony EV-320F*.

Courtesy Sony Corp. of America
(B) *Sony AV-3650*.

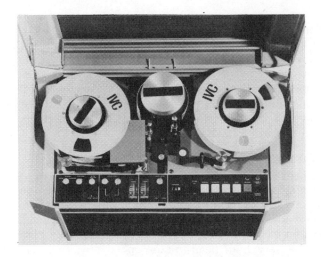

Courtesy International Video Corp.
(C) *IVC-800*.

Fig. 5-1. Examples of helical

(D) *Ampex VR5800.*

Courtesy Ampex Corp.

(E) *Three configurations of IVC-900.*

Courtesy International Video Corp.

video tape recorders.

in this format, since this would require the tape to be wrapped too far around the circumference of a small headwheel.

It was to overcome these problems that the helical tape wrap was developed. Instead of four heads traveling across the tape at almost right angles to the straight path of the tape travel, the tape is now curved around a large drum that contains one or two heads rotating in a plane parallel to the chassis of the machine. The tape is wrapped so that it emerges from the drum at a different level from that at which it entered, describing a helical path as it travels around the drum. In this way, a much longer recorded track is achieved, and this track is at a very sharp angle to the edge of the tape.

By a suitable choice of head-drum diameter and tape width, the track can be made long enough to hold a complete tv field and still maintain a high writing speed. This means that the head-switching and control electronics can be very much simpler than in the quad-head machine. The helical arrangement completely eliminates all head-switching problems and head-matching problems if only one head is used to scan the tape. Alternatively, two heads can be used, with switching occurring at the beginning or end of the field only. In each case, the head rotation is much slower. The new problem introduced, despite these simplifications, is that the tape must now be guided through a very accurate path as it describes the helix and changes height.

In common with the quad-head method, audio and control tracks are recorded in a normal longitudinal manner by separate stationary heads. These heads are placed to one side of the head-drum assembly.

The result of these changes is a machine of very different appearance from the quad-head vtr. It is much simpler and lighter, and its internal electronics and construction are much simpler. Fig. 5-1 shows some typical helical machines.

At this point, it is worth mentioning that another modern development occurred which contributed more than anything else to the realization of a smaller and cheaper vtr. At the time the helical vtr was first introduced, the transistor had become a reliable commercial reality.

It is the helical vtr that has become the standard machine of the nonbroadcast industry, and models and formats abound. However, the basic principles are the same in all machines. In general appearance, they are not unlike any other type of tape recorder. The tape is wound on reels; a supply reel is placed on the left-hand side of the machine (the other side has been used in a few rare instances); the tape passes over a tension arm, in front of an erase head, and around the head drum. After the head drum, the tape passes in front of a control and audio head and then through a cap-stan and pressure-roller assembly before entering the take-up reel.

The main differences from other recorders are the tape guides at each end of the head drum, the drum itself, and the fact that the tape changes level as it passes around the drum. In overall size, helical vtr's are generally a little larger than most audio tape recorders. The manner in which the helical wrap around the drum is achieved varies slightly from machine to machine, as does the placement and form of the pressure roller and capstan used. Some typical wraps are shown in Figs. 5-2, 5-3, and 5-4.

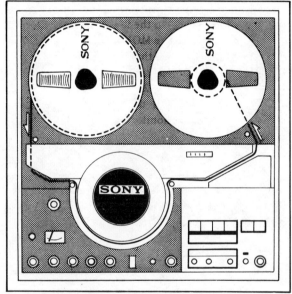

Courtesy Sony Corp. of America

Fig. 5-2. Tape path in Sony EV-320F.

The machine is operated by controls placed on a front panel. These are usually level controls for audio and video, along with a tracking control and a function lever or buttons. If a lever is used, it can be placed in one of several positions: stop, fast forward, re-wind, play, and pause. The record mode is entered by pressing a record button and then moving the lever to the play position. Some of the more expensive machines have a series of push buttons for these functions, and the modes are selected by solenoids when the correct button is pressed. Most machines have a counter that is belt driven from one of the spools. The more expensive machines

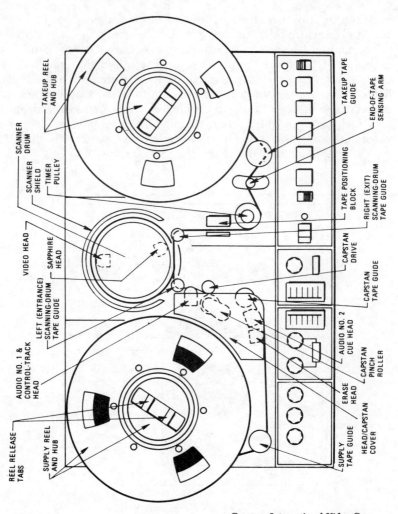

Courtesy International Video Corp.

Fig. 5-3. Tape path in IVC-800.

have a counter driven from the capstan and calibrated in minutes. A few have a motor on-off switch, which will stop the heads while the power is still applied. The various input and output jacks, etc., are mounted on a rear or side panel.

SLANT-TRACK PRINCIPLES

The most vital part of the vtr recording system is the drum-head assembly. Two types of head drum are in use: those with a full

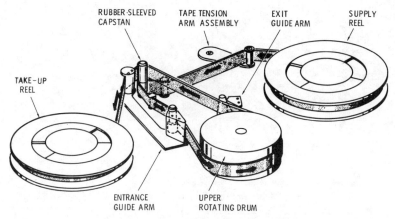

RUBBER-SLEEVED
CAPSTAN

TAPE TENSION
ARM ASSEMBLY

EXIT
GUIDE ARM

SUPPLY
REEL

TAKE-UP
REEL

ENTRANCE
GUIDE ARM

UPPER
ROTATING DRUM

Fig. 5-4. Tape path in Ampex VR5800.

wrap of tape and one record-playback head and those with a half wrap of tape and two heads. The heads are mounted on a disc and rotate in a horizontal plane within the drum, while the tape travels in a slanted path around the outside of the drum (Fig. 5-5).

As the tape is moving slowly around the drum, it has traveled a short distance between the times at which the head starts successive scans. In this way, the head tracks are laid out next to each other in a series of straight, parallel lines across the tape. Fig. 5-6A shows the tape wrapped halfway around the drum. Side A is lowered to the level of side B as the tape describes this half wrap. Point X on side A and point Y on side B are at the same level as the rotating heads, and hence the heads describe a path across the tape from point X to point Y. In this fashion, the path traced by the heads is slanted

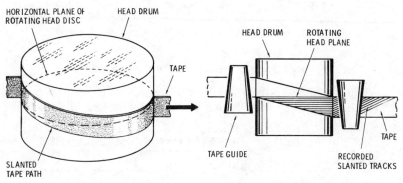

HORIZONTAL PLANE OF
ROTATING HEAD DISC

HEAD DRUM

TAPE

SLANTED
TAPE PATH

TAPE GUIDE

HEAD DRUM

ROTATING
HEAD PLANE

TAPE

RECORDED
SLANTED TRACKS

(A) *Tape path around drum.*   (B) *Recording of slanted tracks.*

Fig. 5-5. Recording of tracks in a helical vtr.

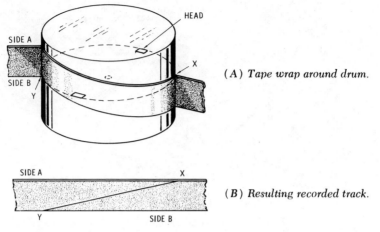

(A) *Tape wrap around drum.*

(B) *Resulting recorded track.*

Fig. 5-6. Slant-track recording.

at a narrow angle to the edge of the tape. When the tape is laid flat as in Fig. 5-6B, the tracks make a sharp angle with the edge of the tape. This angle is determined by the circumference of the head drum and by how far around the drum the tape is wrapped. All of the above is equally true for a full wrap of tape, which produces a longer track across the tape.

Since one complete field is recorded on each track, the time required for the head to travel the length of the track is equal to one tv field, i.e., 1/60 second. Hence, for a full-wrap machine with one head, the drum takes 1/60 second to complete a full revolution, and a two-head half-wrap machine takes 1/30 second.

The time with respect to the tv picture when the head begins and ends its scan must be carefully controlled, and this point is placed in or near the vertical interval. Thus the vertical interval and sync pulses appear near the edge of the tape either at the top or bottom, and not in the center as in the quad-head system.

HEAD AND DRUM ARRANGEMENTS

When the tape wraps around the drum head, the beginning point of the helix can be either higher or lower than the end point (Fig.

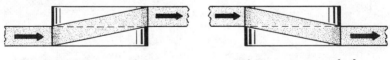

(A) *Beginning point low.*   (B) *Beginning point high.*

Fig. 5-7. Possible tape paths around head drum.

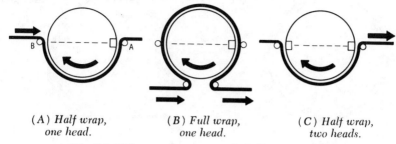

| (A) Half wrap, | (B) Full wrap, | (C) Half wrap, |
| one head. | one head. | two heads. |

Fig. 5-8. Different tape wraps and numbers of heads.

5-7). The Ampex and IVC machines both have the tape wrapped in an ascending helix, but most others use a descending helix.

The choice of the number of heads governs the type of wrap to be used. A one-head machine must have a full wrap of tape. If less than a full wrap is used, there will be too long a gap between the end of one track and the beginning of the next. Consider a half-wrap machine with only one head (Fig. 5-8A); in this format, one field could be recorded on the tape between points A and B, but the next field would be missed entirely as the head traversed from B to A. This idea was used on several early machines and was known as *skip-field* recording. With a full wrap of tape (Fig. 5-8B), the complete successive fields can be recorded with a minimum gap between them.

With two heads mounted at 180°, only half a wrap is required (Fig. 5-8C). In this way the heads can be used alternately with a slight overlap between them by using just over a 180° wrap. This will give full-field recording with no interruption between the successive fields.

Helical machines fall into two distinct groups, those with one head and those with two. A clearer picture of the manner in which the tracks are laid down and of the head-drum assembly required may be gained if the one-head and two-head machines are examined separately.

One-Head, Full-Wrap Machine

In one type of machine, a single video head is used, and the tape is wrapped completely around the drum. The points at which the tape enters and leaves the drum must be very close so that there is a minimum of time when the head is not in contact as it crosses from one side of the tape to the other. Two types of complete wrap can be used, the omega and the alpha (Fig. 5-9).

The alpha wrap has a complete overlapping wrap of tape around the drum, and the guides ensure that the two edges of the tape

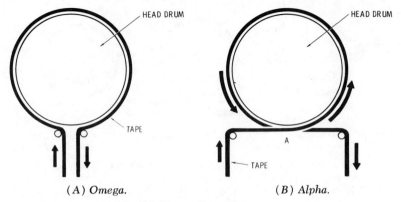

(A) Omega.   (B) Alpha.

Fig. 5-9. Types of complete wrap.

are in contact at the point marked A in Fig. 5-9B. Thus the head never actually leaves the tape as it moves from one side to the other. This affords complete video tracks running across the whole width of the tape, and thus no loss of video information occurs. A slight amount of noise is injected into the picture signal at this crossing, but this is placed in the vertical retrace time and is thus not visible on the screen.

The audio and control tracks are recorded near the edges of the tape and do in fact cut through the video tracks. However, due to their different frequency content, a minimum of interference occurs, and the heads are angled such that even this interference is kept to a minimum.

The omega wrap (Fig. 5-10) is not a complete wrap of tape around the drum. There is a very small section of the drum, between the point at which the tape first makes contact and the point at which it leaves, where the head is not in contact with the tape. This space is kept to a minimum by the tape guides, and the length of time for the head to traverse this gap is so small that no serious picture degradation occurs. This type of wrap allows the heads to begin and end their scan slightly away from the edges of the tape, thus leaving space for the audio and control tracks. Obviously, a small loss of video signal does occur, and on the older Ampex machines this can be seen as a small band of noise a few lines in width at the bottom of the picture. In the later Ampex machines, this dropout was moved into the vertical-blanking interval and lasted for about 10 lines.

Several early Sony machines used a slight variation of this idea. Instead of accepting the dropout, an extra head (Fig. 5-11) was placed in the drum to record and play back the vertical interval only. It was positioned so that when the main video head was tra-

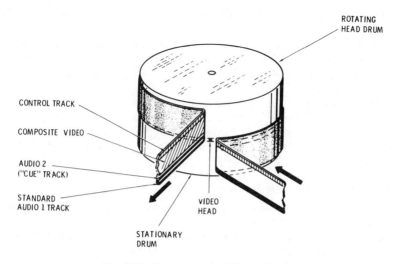

CONTROL TRACK

COMPOSITE VIDEO

AUDIO 2 ("CUE" TRACK)

STANDARD AUDIO 1 TRACK

ROTATING HEAD DRUM

VIDEO HEAD

STATIONARY DRUM

Fig. 5-10. Omega wrap with one head.

versing the gap the sync head was a few degrees away from it. The sync head recorded a short, separate helical track that held all or most of the vertical interval. The heads were switched at the appropriate times. This became known as the "1.5-head system."

The main objections to the full-wrap, one-head format are the dropout period of the signal as the head changes from one side of the tape to the other and the fact that these machines have been

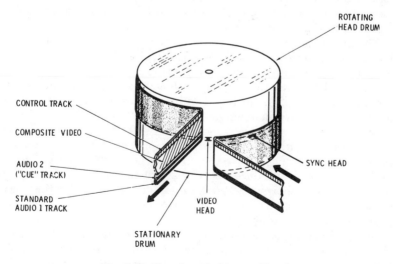

CONTROL TRACK

COMPOSITE VIDEO

AUDIO 2 ("CUE" TRACK)

STANDARD AUDIO 1 TRACK

ROTATING HEAD DRUM

SYNC HEAD

VIDEO HEAD

STATIONARY DRUM

Fig. 5-11. Use of vertical-interval head.

61

prone to "sticktion" (the adhering of the tape to the smooth metal surface of the drum).

The main advantages claimed are that no geometric, mechanical, or electronic head matching is needed, which simplifies the circuitry and head manufacture, and that this type of machine is not subject to color banding.

Because of the potential advantages of this format, it was gradually improved until it was able to compete with the quad system in terms of picture quality and time-base stability. This format is **important because it forms the basis of the broadcast helical machines.**

Two-Head, Half-Wrap Machine

If two heads are mounted in the head drum, they can be used successively in the recording and playback of a tv signal. Since one can be used to take over from the other at some time during the picture, it is possible to wrap the tape less than a full turn around the drum. If a wrap of 180° is used with the heads being switched after equal intervals of time, each head can record a tv field. Thus the head now only has to travel halfway around the drum in 1/60 second, halving the rotating speed from that used in the one-head machine.

In appearance, the pattern of the tracks is much the same as in the full-wrap machine, but there is one important difference. There is no longer a time when no head is in contact with the tape. A wrap of slightly more than 180° allows both heads to contact the tape simultaneously for a brief time, during which the signal can be switched from one to the other. In this way, there is no loss of signal as in the one-head machine.

The main advantages claimed for this type of machine are in its ease of threading and operation, and its freedom from "sticktion," although this has occasionally plagued the 1-inch variety.

The two heads usually come in matched pairs and do not require any operator or field adjustments other than a minor electronic adjustment to achieve matching.

Fig. 5-12 shows two heads mounted at 180° on a rotating disc in a head drum, with the tape wrapped just over halfway around. Note that the direction of rotation of the head disc is in either direction. The tracks produced by this arrangement are shown in Fig. 5-13. Fig. 5-14 shows the tracks in more detail, indicating how adjacent tracks are recorded by alternate heads. The horizontal-sync pulses are aligned as shown.

Fig. 5-15 shows the relationship between the tracks and the video signal recorded on the tracks. Note the overlap position where the heads are switched. This is in the last few lines of video before the

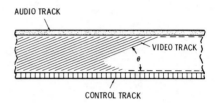

Fig. 5-12. Two heads with half wrap of tape around drum.

VIDEO HEADS

HEADS ROTATE IN EITHER DIRECTION

TAPERED GUIDE

DIRECTION OF TAPE TRAVEL

Fig. 5-13. Tracks on video tape.

AUDIO TRACK

VIDEO TRACK

θ

CONTROL TRACK

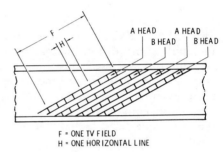

Fig. 5-14. Details of video tracks.

F

H

A HEAD   A HEAD

B HEAD   B HEAD

F = ONE TV FIELD
H = ONE HORIZONTAL LINE

NOTE: NOT TO SCALE

vertical interval. This ensures that switching transients will not upset the vertical-sync circuits, and in most tv sets the switch point cannot be seen on the screen.

On playback, head A does not necessarily have to scan the same track that it recorded, and neither does head B. Good head matching in manufacture will ensure the recording of identical tracks that play back equally well on either head.

Comparison of Full-Wrap and Half-Wrap Machines

Although there are several differences between the two types of helical machine, there are many items and problems that are common to both. Although the actual head assembly differs greatly from that of the quad-head machine and the two types have some differences, several important similarities exist.

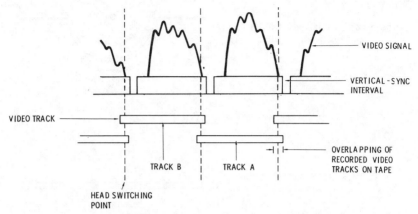

Fig. 5-15. Relationship between tracks and video signal.

In both types, the drum is a cylinder about six inches in diameter with a horizontal slit about halfway up. The rotating heads are mounted in the drum and protrude a short distance through the slit. The rotating mechanism inside the drum is a plate on which the heads are mounted, and the plate, or disc, is driven by an electric motor mounted beneath the drum assembly and inside the machine. In some machines, the entire upper half of the head drum rotates, and the single head is mounted on the circumference.

All helical machines have a much larger head-drum mass than the quad-head machine has, and hence the head-servo problems are increased to a degree. However, this is offset by the longer track and the larger amount of video information in the track.

The two-head half-wrap types became the most popular for industrial and educational use. The one-head type was eventually developed for broadcast work.

TAPE GUIDES

To achieve the slanting path around the head drum, the tape is passed around finely machined guide posts located both before and after the drum. These guides are angled or are cone shaped, and their effect is to change the direction of the tape travel with respect to its width (Fig. 5-16). This allows the tape to leave a horizontal reel, be angled up or down, and then be made horizontal again on a higher or lower plane. Although the guides in different machines may have differing physical construction and appearance, they all

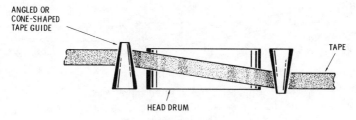

ANGLED OR CONE-SHAPED TAPE GUIDE

TAPE

HEAD DRUM

Fig. 5-16. Use of tape guides.

accomplish the same purposes. Briefly the main purposes of the guides are:

1. To guide the tape around the circumference of the drum for an angle of 180° or 360°, depending on the type of machine.
2. To give the tape a precise upward or downward slant as it passes around the drum. The angle of this slant is critical because it establishes the correct helix path and the correct angle of the tracks. Since the tracks are between 5 and 10 mils wide and have a guard band of 3 to 6 mils, it is obvious that the guides must be very accurately aligned.
3. To ensure that the tape enters and leaves the drum assembly in a precise spatial relation to the rotating heads.
4. To keep the tape absolutely flush against the surface of the drum and to maintain precise tape penetration by the heads.

Guiding the tape in a helical path requires extremely precise machining and adjusting of the parts. The tape guides on a helical machine are very delicate pieces of machining that do not have an exact counterpart in any other type of magnetic recorder. They must be exactly the same in all machines of the same type, or interchange cannot be achieved. Interchange is the ability to play any tape on any machine of the same type, and it is a most important feature of vtr's.

The guides before and after the tapered guides are straight vertical posts with narrowed sections equal in width to the width of the tape. These are set at exact heights to feed the tape to the tapered guides and the drum at the correct height for proper scanning. Like the tapered guides, they should never be adjusted after the machine is set up.

To lessen the angle through which the tape must bend, some machines have used a feed reel mounted on a slanted table (Fig. 5-17) or have used an inclined cylinder as part of the head-drum assembly (Fig. 5-18). Whichever method is used, the heads still rotate in a horizontal plane.

Fig. 5-17. Use of tilted supply reel.

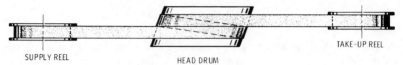

Fig. 5-18. Use of angled head drum.

Between the two tapered guides, the tape remains in contact with the surface of the drum. Its height with respect to the rotating heads is controlled by one of three methods:

1. A guide band running the complete tape bath around the drum. This is used on the half-wrap machines only.
2. Eccentric washers at each end of the tape travel.
3. An eccentric screw at the halfway point around the drum. This is used exclusively in the 1-inch, full-wrap machines.

These methods are illustrated in Fig. 5-19.

The most important point to observe with respect to eccentric washers or setscrews is that they have been factory set and then

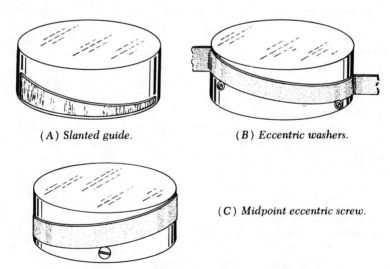

(A) Slanted guide.   (B) Eccentric washers.

(C) Midpoint eccentric screw.

Fig. 5-19. Methods of guiding tape around drum.

tightened or painted in place. They should never be moved or dis-turbed, because they set the correct height of the tape at a crucial part of its path. This setting is so critical that a small build-up of tape oxide can cause mistracking and give serious interchange problems. After every few passes of tape, the screws or washers should be thoroughly cleaned.

The apparent simplicity of the tape guides belies the amount of research and engineering that has gone into their development. Many of the interchange problems in the early machines were the result of tape-guide trouble, and the path to the modern reliable vtr has not been easy. Tape guides should be treated with the same respect and care that is given the rotating heads. They are just as im-portant and just as expensive, but not as easy to change or adjust. Due to the accurate machining and manufacture of the guides, the only thing a machine operator should do is to keep them scrupu-lously clean, and they should never be moved or tampered with.

THE TAPE DECK

In addition to a head drum and tape guides, a vtr must have other facilities. It must be able to record and play back audio and control tracks, and it must have one or more erase heads for both the video and audio tracks. A typical layout is shown in Fig. 5-20.

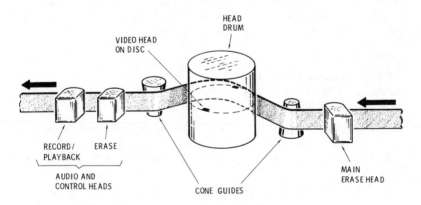

Fig. 5-20. Arrangement of heads in a helical vtr.

The video tracks do not run completely across the tape from edge to edge; a guard band is left between the video tracks and the other tracks. The width and exact placing of these tracks varies from machine to machine, and no standard exists among the various manufacturers. This is one reason why a tape made on one machine will not play properly on a machine of a different type. The full-width erase head is always placed at the tape-entry side of the drum, and the audio and control heads are placed on the other side. Often the audio and control heads are contained in the same shielded housing.

THE CAPSTAN ASSEMBLY

A constant-speed drive is applied to the tape by a capstan and pressure-roller arrangement. This is similar to the operation of audio and other recorders. It is covered in more detail later.

ELECTRONICS

The electronics of helical vtr's can be divided into two main sections, the servos and the record and playback circuits. The electronics is essentially the same in both the one- and two-head machines, the only major difference being the incorporation of head switching in the two-head models. All of the electronics in the helical machine is much simpler than in the quad-head machine. In general, only the barest functions are provided for, and they are performed in a simple, straightforward manner. The minimum requirements are obviously recording, playback, and servo control of the head. Electronics-to-electronics (E-E) operation is also provided, since this requires no extra circuitry but merely the simultaneous use of the recording and playback chains.

Servos

As in the quad-head machines, helical vtr's must have a servo-mechanism to control the tape and the heads. Basically the servo must accomplish three things:

1. Place the heads in the correct position with respect to the tape and the tv signal to record the signal in a definite pattern.
2. Effect correct tracking of the tape and heads in the playback mode.
3. Effect head switching on both record and playback.

To perform these functions, the servo requires a signal that indicates the position and speed of the heads. For this, coils and metal vanes are mounted in the head-drum assembly, and a controlled electric motor or braking system is mounted under the head-drum assembly. The servo circuit needs some pulse-shaping circuits to handle the pulses from the control track and the head coils, a pulse record amplifier for the control track, a motor drive amplifier and a braking system for the head motor, and a servo comparator circuit. The manner in which these are made to perform their functions is left until a later chapter devoted entirely to servos.

Head Position—In the longer-tracked helical format, one complete field is recorded on each track. In this case, it is necessary to decide where the field is to begin and where it is to end, and also where to change over to the next field. This in effect means deciding on a place on the tape for the vertical interval to the placed. In the helical format, the usual place is near one end of the track instead of in the middle as in the quad-head format. Whatever point is chosen, it obviously must be adhered to for all fields and all tracks, and this means the rotating head must assume a definite position with respect to the vertical interval. These considerations apply to both full- and half-wrap machines.

In order to place the vertical interval at the top of the track, the head must be in the position shown in Fig. 5-21. The head comes into contact with the tape about 8 or 10 lines before the vertical interval and is switched to the circuit about 3 or 5 lines before the vertical interval. Then the head continues to scan until it leaves the tape at the other edge at the occurrence of the next vertical interval. In this manner, the tracks are laid down across the tape in an orderly fashion. They are parallel to each other and are separated by a guard band of definite width.

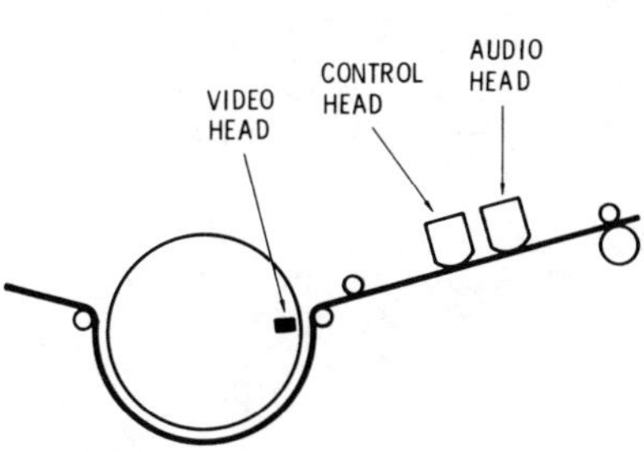

Fig. 5-21. Head position for placement of vertical interval.

Tracking—On record, the head tracks are laid down as shown in Fig. 5-22A. On playback, the heads obviously must follow the same path, as shown by the dash line at A in Fig. 5-22B. If the head follows path B, reduced video output will occur. Path C will give little or no output, depending on the width of the track and the guard band. Path D shows what happens when the head crosses more than one track; this produces noise bands in the picture. The servo has to produce condition A and prevent B, C, and D. These last three conditions are known as *mistracking*.

The distance between the tracks is of importance. If the guard band is too narrow, it will be easy for the head to encroach upon the wrong track, particularly if the machine is slightly misaligned.

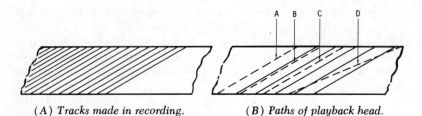

(A) *Tracks made in recording.*   (B) *Paths of playback head.*

Fig. 5-22. Tracking in a helical vtr.

The wider the guard band, the less is the chance of serious mis-tracking.

As each vertical-sync pulse occurs, a pulse is recorded on a track along the edge of the tape. If the tape is running at an even speed, these pulses will be evenly spaced along the tape. On playback, they produce regular pulses that are fed into the servo electronics and used to monitor and correct for tape-transport errors. The occurrence of these pulses is used to indicate when the vertical interval on the tape is in the correct position for correct playback by the heads. The position of the heads has to be monitored and controlled, and for this, coils and rotating metal vanes are mounted inside the head drum.

Refer to Fig. 5-23. On record, the point marked X on the tape, the point marked Y on the drum, and the head were coincident. This is where the head began its scan across the tape at the start of a new tv field. On playback, these three points must again align with

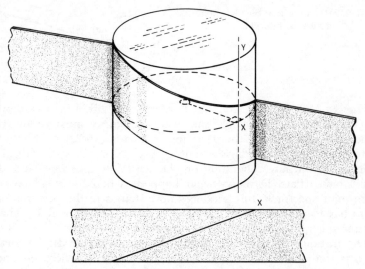

Fig. 5-23. Alignment required for correct playback scanning.

the same precision they did on record. This three-way alignment must be repeated after the head has completed one revolution and the tape has moved linearly to bring the next track into position. If this alignment is maintained, correct playback scanning of the tape occurs, and a stable picture is played back. If it is upset, mistracking is observed.

Record and Playback Electronics

The recording chain consists of an input video amplifier, an fm modulator, and an amplifier for the record head. A minimum number of controls and adjustments are needed, so the complete recording chain is very simple. The incoming video is amplified, applied to the fm modulator, and fed to the record heads through the head amplifiers.

For playback, a head preamplifier takes the very small signal from the heads and amplifies it to a level suitable for feeding to a series of fm limiters. These remove all the amplitude variations and thus eliminate most of the noise. After this, the signal is demodulated and passed through a video output amplifier.

Electronics-to-electronics (E-E) is provided in the record mode. This is the method whereby the incoming signal is viewed on a monitor at the output of the machine as it is being recorded. The signal is fed from the fm modulator to the limiters at the same time it is fed to the record heads. In this way, the signal passes through all of the record and playback electronics simultaneously and provides a good checking procedure for the operation of the machine. It checks everything except the tape itself; this must wait until the recording can be played back.

The sync is stripped from the incoming video, and the vertical interval is separated. It is used with a pulse derived from the rotating head mechanism to control the head servo and to record a control track along one edge of the tape. On playback, the control track is used instead of the vertical interval to control the head servo.

The audio section is no different from an ordinary audio recorder. It contains a simple record-playback head and amplifier. The bias-erase oscillator is used to feed both the main erase head and the audio erase head.

One-Head, Full-Wrap Machine—Fig. 5-24 shows a block diagram of the electronics in a typical one-head machine. The head motor can be either the same one used to drive the capstan or a separate motor.

The placement of the dropout time is shown in Fig. 5-25. It can be either just before the vertical interval or in the back porch of the interval. It does not matter which place is used, provided the dropout does not interfere with either the vertical-sync pulse or the pic-

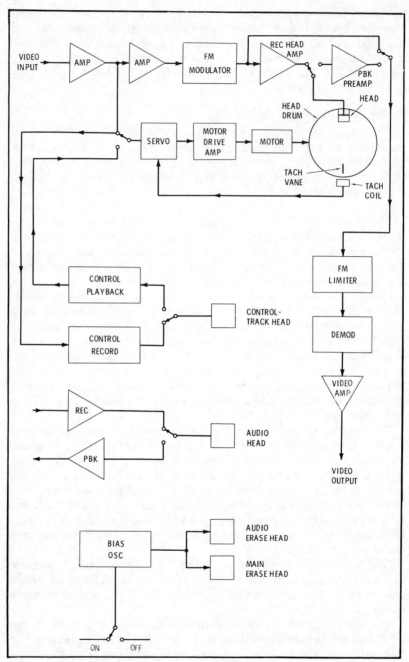

Fig. 5-24. Block diagram of one-head machine in record mode.

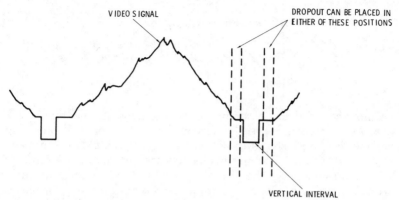

Fig. 5-25. Placement of dropout time of one-head machine.

ture information. Maintaining its position is the function of the servo control circuitry.

Two-Head, Half-Wrap Machine—The block diagram of the electronics in a two-head machine is similar in principle to that of a one-head machine, with the addition of head-switching circuitry and an extra tach sensor in the head drum. Fig. 5-26 shows the added sections.

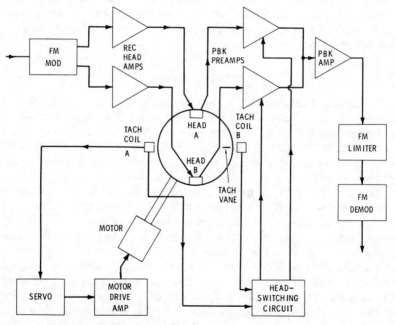

Fig. 5-26. Partial block diagram of electronics in two-head machine.

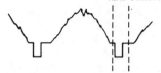

Fig. 5-27. Head-switching points for two-head machine.

Although the drum assembly differs somewhat, the tape track layout is not significantly different from that of the one-head machine. Fig. 5-27 shows the head-switching points. Note that the positions at which switching is effected are just about the same as those points chosen for the dropout period in the one-head machine.

Head Switching—Head switching is required in the two-head machines only. The changeover from one head to the other is effected on both record and playback by the same method. The coils in the head drum that produce the pulses for the servo are also used for this purpose. The pulse is fed to a switching circuit that transfers the signal from one head to the next. The circuitry is arranged so that the head scanning the tape is always the active head.

In some machines, the signal was fed continuously to both heads on record, but on playback the nonscanning head was switched off to prevent noise. The earlier machines did not have electronic switching, but used a split ring and head brushes to transfer the signal. Whichever method is used, switching must occur at the correct time in the tv signal and when the heads are at the correct physical place on the head drum.

CONCLUSION

The greatest diversity of decision in the design of a helical vtr involves the head drum, tape speed, tape width, the number of heads to be used, and the type of wrap. Many of the governing factors are not independent, they are often contradictory, and a clear-cut decision on many things is impossible. The resultant vtr is thus a compromise of conflicts and an exercise in difficult decision.

The first thing to be decided is the number of heads. Although the first helical machine had two heads and a full wrap of tape, only half of this wrap was used to scan the tape. This was an odd format which was shortly replaced by the same manufacturer with a machine of one-head, full-wrap design.

The decision on the number of heads actually determines the effective amount of wrap, and thus determines the type of tape wrap to be used. The type of wrap affects the length of the recorded tracks; the longest tracks are obtained with a full wrap. A longer track to be scanned in a given time naturally means a higher writing speed and thus a better signal and higher-quality picture. Although the number of heads and type of wrap do not affect the width of the tape, the width of the tape does affect the length of the track. The wider the tape is, the longer the track is.

The best possible combination would appear to be a full wrap with a very wide tape. The widest tape available for vtr's is 2 inches, but this is very expensive and requires excellent mechanics to control its path and tracking. So a 2-inch tape increases the cost of the machines as well as the tape inventory.

With a full wrap of very wide tape, another serious problem arises, that of "sticktion" (the adhering of the smooth surface of the tape to the smooth metal of the head drum). This problem increases with a highly polished tape and drum and is also highly dependent on temperature and humidity. It can be lessened if not completely eliminated by the use of narrower tape and less than a full wrap. Hence, it would seem a narrow tape would be expedient. For nonbroadcast purposes, adequate performance can be obtained with both one-inch and half-inch tapes.

Helical machines all have a much larger head-drum circumference and a much larger rotating head mass than is found in quad machines. This means that the servo control is much more imprecise. Consequently, helical vtr's do not have the same horizontal stability as the quad-head type, and usually they are not up to broadcast specifications. Those which do have this stability and have a bandwidth sufficient to produce a good quality picture contain much more complex electronics than the simple machines, and this is reflected in their cost and physical size.

There are several ways of classifying helical vtr's, the most common being by tape width, number of heads, and method of wrap. Often all three are quoted in the description of a machine. One or two heads, full or half wrap, and one-inch or half-inch tape may be used. Although all of these options theoretically can be mixed at will, only certain combinations have appeared and are likely to do so in the future. All one-head machines use a full wrap, and all two-head machines use a half wrap. All the one-head machines happen to use one-inch tape, but the two-head machines use both one-inch and half-inch tape.

The chapters that follow describe in detail the construction of the various sections of the helical vtr. There is some interdependence between the sections, and therefore a small amount of repetition occurs, but mainly the discussions are separate and distinct in their content. The information they contain reflects current practice and is taken mostly from the most readily available service manuals of those machines which are in widespread use. For the remainder of this book, only those machines appearing in practice will be covered.

6

The Mechanics of
Helical VTRs

In an audio recorder, the tape is moved past a series of fixed heads by having a constant-speed electric motor drive a capstan and pinch roller. In a vtr, the same kind of motion is required, and it can be accomplished in the same way, but the precision of the path which the tape follows is much more critical. The tape must follow an exact path on record, and it must repeat this path exactly on playback. In playback, the vtr has one requirement not found in any other type of recorder: the tape must be in the same relative position with respect to the video heads and the tv signal as it was on record. Not only must the tape keep to this exact path and travel at an exact speed, but the heads must maintain an exact spatial relationship with the tape and an exact time relationship with the tv signal.

To effect this requires some excellent mechanical engineering as well as some complex electronics. This is further complicated by the fact that all vtr's of the same type must be made to exactly the same dimensions, or interchange is not possible.

The tape deck has to support the head drum as well as all the other heads, tape guides, reels, motors, and other necessary items. The operational demands on a modern vtr are such that it should be capable of being used by anyone with a minimum of training, and that maintenance should be relatively easy to perform. The following sections show the main points of mechanical construction that lead to such a machine.

HEAD-DRUM ASSEMBLY

No matter what type of machine is contemplated, the head-drum assembly probably involves the greatest mechanical decision that has to be made. It is here that the highest in mechanical engineering has to be achieved, with a repeatability of parts unequalled in most other spheres of engineering. The head drum assembly has to hold the rotating heads so that they maintain a correct rotating plane, and it must support and guide the tape around it in the correct path. The heads must rotate in an exact circle between the upper and lower halves of the drum, regardless of whether one half rotates or not. The two halves of the drum must be exactly concentric so that the tape is held in constant contact with both halves throughout its path, and the head must be rotated in an exactly concentric path with a specified slightly larger diameter than the drum in order to produce an exact amount of tip protrusion into the tape.

In addition to accurate positioning of the tape, the heads must be precisely located with respect to the position of the tape. Thus the height of the rotating platform or disc on which the heads are mounted must be very accurately determined. When mounted in place, the heads must scan an exact horizontal plane above a reference position. This must be exactly the same on all machines of the same type, and it must be preserved when the heads are changed. A 5-mil deviation on a head 10 mils wide will lose half the track on a prerecorded tape. If the tracks are placed close together, this will cause the head to pick up low-level signals from two tracks simultaneously. The bearings that support the main central spindle are set in place with extreme precision to ensure that both halves of the drum and the heads are exactly concentric with the rotating axis.

The head-drum assembly must also support the various tach coils or photoelectric devices used to sense the head rotation, and sometimes the head preamplifier circuitry is also mounted in the head drum. The underside of the drum is used for mounting the whole assembly to the chassis of the machine, and the main central drive spindle projects from beneath the unit; a drive wheel is mounted to the spindle so that it can connect to the head-drum motor. Fig. 6-1 is typical of the diagrams in service manuals that illustrate this construction. The various drive and control methods are covered elsewhere, since these do not involve physical connection to the assembly.

It is possible to remove the whole unit from the machine, but this is done only for major repairs such as bearing replacement or the renewal of the entire drum assembly. Otherwise, the only mechanical work done on the assembly is to change the heads or the tach sensors.

In most machines, the drum is mounted with its cylinder sides

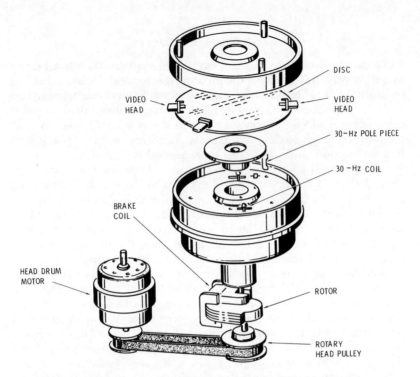

Fig. 6-1. Partially exploded view of typical head-drum assembly.

vertical and the head rotating in a horizontal plane. The tape, on being wrapped around the drum, changes the height of its vertical plane in order to produce the helix. It does not matter whether the tape lowers or raises its level; both systems are used.

Usually the reels are mounted horizontally, but to ease the guiding of the tape as it begins its helix, occasionally one of the reels may be tilted. A few machines use an angled head-drum assembly with the top truncated in a horizontal plane. In this type of machine, the tape continually slopes around the assembly in its helical path so that the head, which still rotates in a horizontal plane, will describe the normal slanted tracks. These remarks apply to both full- and half-wrap machines.

The tape guides are part of the drum assembly and are mounted and built as an integral part.

To prevent the tape from sticking to the drum, which it will do because of the large area of contact and the tension, air is forced up through the head assembly and out through the slit by a fan mounted on the bottom of the central shaft. This forms an air bearing that effectively lubricates the tape and the drum to prevent "sticktion." The air is also used for cooling inside the machine, and care must be taken in mounting not to obstruct the air intake. This will cause over-heating of the motors and bearings as well as "sticktion" problems. A few machines are equipped with a heat sensor that disables the power under these conditions.

Figs. 6-2 and 6-3 show typical head-drum construction. Note the position of the tape guides and guide band on the outside of the drum (Fig. 6-2), the tach coils inside the head (Fig. 6-3), and the frequency-generator coil and toothed wheel beneath the head (Fig. 6-3).

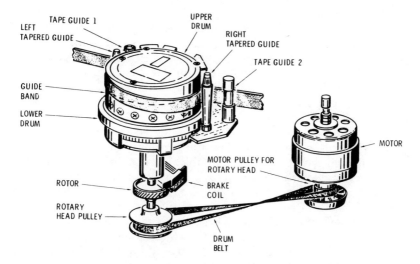

Fig. 6-2. Construction of typical head drum.

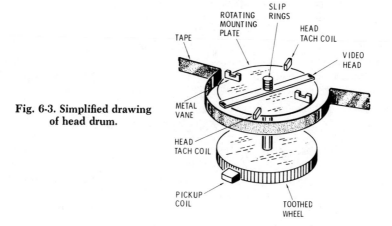

Fig. 6-3. Simplified drawing of head drum.

SMOOTH SURFACES

The metal surface of the drum over which the tape moves is very smooth, as is the oxide surface of the tape. Pits and bumps are far too small to be observed by eye or felt by hand, as they are of microscopic size. However, a degree of roughness does exist. Continual polishing of two metal surfaces can produce a smoothness so fine that if the two are brought together, the intermolecular attraction is so great that they can only be separated with extreme difficulty; and if the separation is not effected properly, the smoothness of the surfaces will be impaired.

After much use, this effect becomes apparent between the head drum and the tape in a helical vtr. It is known as "sticktion." Its effect is to slow the tape motion and sometimes even stop it altogether. In order to prevent this, a new head drum on a vtr is not polished to mirror smoothness and is quite dull in appearance. The acquiring of a mirror-like surface after much use is an indication of wear and a sign of impending trouble.

Some modern tapes have the oxide side so smoothly made and polished that it is more shiny than the backing. This type of tape on an old machine will often give trouble. Various other factors, such as tape characteristics, temperature, humidity, etc., will make this type of trouble very intermittent; also, some machines just do not "like" some tapes because of their relative smoothness.

If an old machine has a shiny head drum, it can be roughed up by lightly rubbing with crocus cloth. The cloth should be rubbed in the direction of the tape travel only, not up and down. If this procedure is attempted, cover all of the drum exposed to tape and do not touch the head or guides. Afterward, clean all the area around the head thoroughly to remove excess dust.

This action will sometimes improve the situation, but it is not a guaranteed procedure. Often the only safe way to eliminate the problem in an old machine is to install a new head drum. With a new machine, the best method is to use a recommended tape. As far as possible, everything should be kept in the same air-conditioned environment.

THE HEAD DRUM AND THE RECORDED TRACKS

The dimensions of the head drum and the tracks on the tape are dependent on many interrelating factors. The first decision to be made is the highest playback cutoff frequency, which is determined by the electronic circuit of the fm modulator. The next factor is the minimum head gap that can be manufactured or delivered reliably. After these have been decided, the necessary head-to-tape speed, or scanning speed as it is sometimes called, can be determined. The exact calculations and procedures differ from manufacturer to manufacturer, but the principle is outlined in the next few sections.

Head-to-Tape, or Scanning, Speed

The scanning speed is dependent on the smallest possible wavelength the heads can handle on playback. It is given by

$$v = 2GF$$

where,
  v is the scanning speed,
  G is the head gap,
  F is the maximum playback frequency.

These are often not exact figures, and the speed becomes a rounded-off or buffer figure, chosen for convenience to handle the available head gap and several fm frequencies. It can also be altered to accommodate other difficult mechanical problems.

Track Length

The track length is determined from the scanning speed and the fact that one complete tv field is recorded on each track. This leads to the simple formula

$$TL = \frac{v}{60}$$

where,
  TL is the track length,
  v is the scanning speed.

For several reasons, this figure is not the same as the drum circumference; this is covered in a later section.

Track Angle

The angle (θ) the tracks make with the edge of the tape is easily determined from

$$\sin \theta = \frac{\text{Used width of tape}}{\text{Track length}}$$

Several factors can vary this angle slightly. In most machines, the video tracks do not scan the whole width of the tape, which lessens the angle slightly.

Horizontal-Sync Pulse Alignment

Horizontal-sync pulse alignment is required only in a machine that incorporates slow speed and still frame. It is first necessary to know

81

the distance between the successive horizontal pulses as they are recorded. With 262.5 lines per field, this distance is

$$H = \frac{\text{Track length}}{262.5}$$

If one track begins with a full horizontal line of video, it will end with a half line, and the next track will begin with a half line and end with a full line (Fig. 6-4).

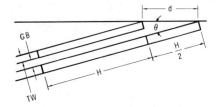

Fig. 6-4. Horizontal-sync pulse positions on successive tracks.

Guard Band and Video Track Widths

To lay out the tracks so that horizontal alignment occurs, the track separation and the tape speed must both be chosen properly. It is best to fix the distance between the tracks first, because there are several separations that can be used, and the minimum that will not produce overlapping tracks must be chosen for tape economy. To find the distance between the tracks, Fig. 6-4 is used.

$$\tan \theta = \frac{\text{TW} + \text{GB}}{\frac{1}{2}\text{H}} = \frac{2\text{T}}{\text{H}}$$

$$d = \frac{\text{T}}{\sin \theta}$$

where
TW is the track width,
GB is the guard-band width,
H is the horizontal-pulse spacing,
T is TW + GB,
d is the linear distance the tape moves in one tv field.

The track width is determined by the construction of the head, and the minimum figure for T that can satisfy the above equation is likely to be less than the track width. Consequently, the equation must be reworked using 3H/2, 5H/2, etc. (Fig. 6-5), until a figure is produced that will give a minimum separation. Once this final figure for T is found, the guard band can be determined.

Linear Tape Speed

Once the guard band has been determined, the linear tape speed can be determined easily. In Fig. 6-5, the distance the tape travels in one tv field is d. In one second, 60 of these distances are covered; thus the linear speed is given by

$$S = 60d$$

It is, of course, possible to select the linear tape speed first, check that it provides the track separation, and then adjust it if necessary to give the horizontal-sync pulse lineup. Most manufacturers' litera-

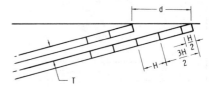

Fig. 6-5. Determination of guard band between tracks.

ture would suggest that this was the procedure. Several tape speeds will result in this alignment, but obviously the track spacing will change. The wider the track spacing, the less chance there is of the head crossing adjacent tracks when minor mistracking occurs. This is an advantage in picture playback.

Drum Circumference

At first, it might appear that the drum circumference is the same as the track length in a one-head machine, and twice this figure in a two-head machine. However, this is not exactly true; there are a few factors that tend to alter this figure.

In the case of a complete alpha wrap with one head, the track length and drum circumference would seem to be an exact match, but this assumption does not take into account the fact that the tape moves a slight distance in the time required for the head to complete one revolution. This has the effect of making the track a little longer than the drum circumference.

With an omega wrap and one head, considerable latitude can exist, depending on how much of the drum is not in contact with the tape at the entry and exit points.

With a half-wrap, two-head machine, the actual track length is governed by the amount of overlap between the tracks, and this must be taken into account before the drum dimensions can be determined.

Although the drum measurements are of critical importance in the manufacture of the vtr, their exact sizes are a matter for individual calculation with each designer within each format, and they do not lend themselves to a ready formula.

TAPE FORMATS

To illustrate the difference in tape formats possible, two 1-inch formats and the most widely used ½-inch format are shown.

Table 6-1. Specifications of IVC VTRs

Parameter	Specification
Scanning Speed	723.18 in/s
Drum Circumference	11.938 in
Track Length	12.053 in
Drum Diameter	3.800 in
Track Angle	4° 45′
Track Width	6 mils
Guard-Band Width	3.5 mils
Tape Linear Speed	6.91 in/s

The IVC Format

Table 6-1 shows the specifications of the IVC machines. Fig. 6-6 shows the track configurations on the tape.

The Ampex Format

Table 6-2 shows the published details of the Ampex format. Fig. 6-7 shows the layout of the tracks. (Linear dimensions are in inches.)

The EIAJ-1 Format

A few years ago, the Electronic Industries Association of Japan decided to prevent further format chaos with monochrome vtr's produced by the Japanese manufacturers. This was in some large measure reinforced by the Japanese government, which demanded a standard format of ½-inch vtr that could be used interchangeably on a national level in the schools. The result is the EIAJ-1 standard. The major points of interest of this standard are given here.

A 12.7-mm (½-inch) tape width is specified, with a two-head slant-track recording system. The main dimensions of interest are

Table 6-2. Specifications of Ampex VTRs

Parameter	Specification
Scanning Speed	1000 in/s
Drum Circumference	16.67 in
Track Length	16.67 in
Drum Diameter	5.3 in
Track Angle	3° 6′
Track Width	6 mils
Guard-Band Width	2.7 mils
Tape Linear Speed	9.63 in/s

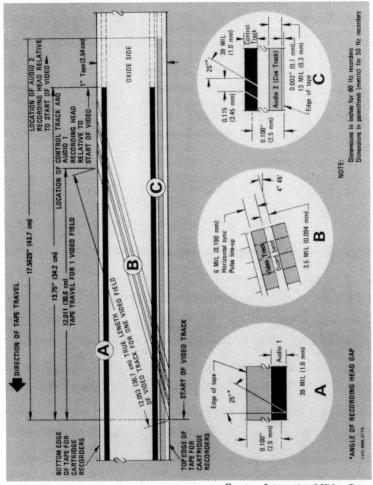

Courtesy International Video Corp.

Fig. 6-6. Tape format used in IVC vtr's.

listed in Table 6-3 and Fig. 6-8. The EIAJ-1 specifications also contain dimensions concerning tape guides, reel sizes, tape tension, etc., which present a complete set of mechanical details for the construction of this type of vtr.

Also, of course, the electronic parameters are just as accurately specified. Fig. 6-9 indicates the frequencies required for the fm modulator. Although the lowest frequency specified for sync tips is "greater than 2 MHz," most manufacturers use 3.0 to 3.2 MHz. The total fm spread is given as 1.6 MHz; thus the top frequencies encountered approach 5 MHz.

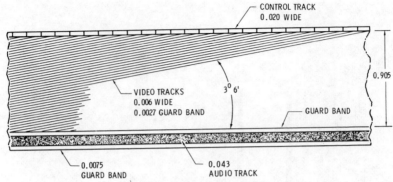

Fig. 6-7. Tape format used in Ampex vtr's.

The position of the changeover from one head to the next is also detailed. Its position in the video signal is shown in Fig. 6-10.

It should be understood that this is not a full set of the EIAJ-1 specifications. These can be obtained from the Electronic Industries Association of Japan if required.

HEADS

The heads are mounted so that they rotate in a horizontal plane and protrude a few mils through a slit in the head-drum cylinder. Various mounting techniques are used for both the drum and the heads; a common basic form is shown in Fig. 6-11. The drum is split into two horizontal sections by a slit that runs around its entire circumference. The lower half of the drum is fixed to the chassis of the

Table 6-3. EIAJ-1 Specifications

Parameter	Specification	
	Metric	English
Scanning Speed	11.1 m/s	437 in/s
Drum Circumference	363.58 mm	14.32 in
Drum Diameter	115.82 mm	4.56 in
Track Length (Approx)	185.3 mm	7.29 in
Track Angle (Tape Moving)	3° 7' 43"	3° 7' 43"
Track Angle (Tape Stationary)	3° 11'	3° 11'
Track Width (Minimum)	0.1 mm	0.003937 in
Track Guard-Band Width (Min)	0.04 mm	0.001575 in
*Track Pitch	0.173 mm	0.0068 in
Linear Tape Speed	190.5 mm/s	7.5 in/s

* Track pitch means the distance between the reference edges of the video track. It is an accurate measurement due to the mechanical construction of the tape guides and the scanning head. The width of the video track can vary due to head construction and wear, which will also vary the guard-band width. Hence the minimum specified track and guard-band widths do not total the track pitch.

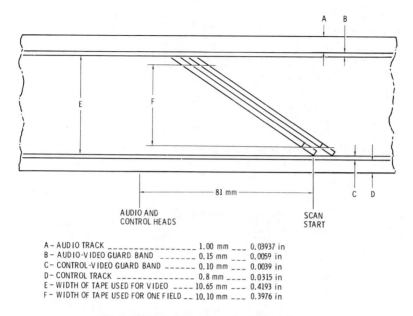

A – AUDIO TRACK _____ 1.00 mm ___ 0.03937 in
B – AUDIO-VIDEO GUARD BAND _____ 0.15 mm ___ 0.0059 in
C – CONTROL-VIDEO GUARD BAND _____ 0.10 mm ___ 0.0039 in
D – CONTROL TRACK _____ 0.8 mm ____ 0.0315 in
E – WIDTH OF TAPE USED FOR VIDEO ____ 10.65 mm ___ 0.4193 in
F – WIDTH OF TAPE USED FOR ONE FIELD __ 10.10 mm ___ 0.3976 in

Fig. 6-8. Track dimensions for EIAJ-1 format.

machine. Inside this half is a horizontal plate attached to a central shaft that is driven by the head motor.

The heads are mounted on the plate in a variety of ways. The three most common are:

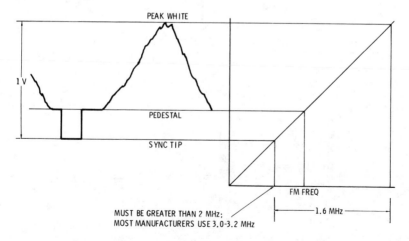

Fig. 6-9. Recorded frequencies for EIAJ-1 standard.

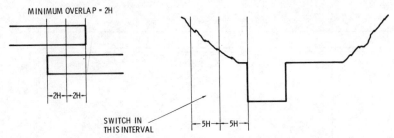

MINIMUM OVERLAP = 2H

SWITCH IN THIS INTERVAL

Fig. 6-10. Head-switching interval for EIAJ-1 standard.

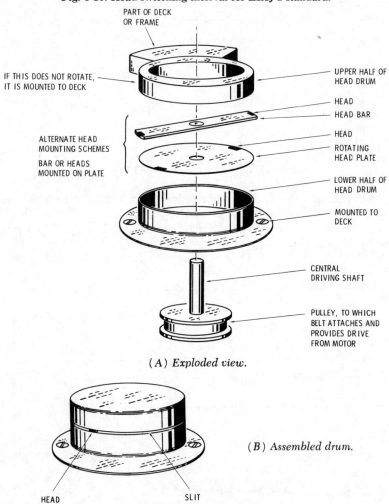

PART OF DECK OR FRAME

IF THIS DOES NOT ROTATE, IT IS MOUNTED TO DECK

ALTERNATE HEAD MOUNTING SCHEMES

BAR OR HEADS MOUNTED ON PLATE

UPPER HALF OF HEAD DRUM

HEAD

HEAD BAR

HEAD

ROTATING HEAD PLATE

LOWER HALF OF HEAD DRUM

MOUNTED TO DECK

CENTRAL DRIVING SHAFT

PULLEY, TO WHICH BELT ATTACHES AND PROVIDES DRIVE FROM MOTOR

(A) *Exploded view.*

(B) *Assembled drum.*

HEAD

SLIT

Fig. 6-11. Basic construction of head drum.

1. The heads are premounted on a bar that is bolted to the plate. Predrilled tapped holes provide accurate location, and this permits two heads to be changed easily and quickly.
2. The heads are lowered into a shaped indentation in the plate. This allows quick, easy replacement of the head, and the mounting is such that the height is set and the penetration can be adjusted easily.
3. The head is slotted into the indentation in the plate by pushing it through the scanning slit.

Whichever method is used is so arranged that head changing does not upset the critical alignment of the rest of the head-drum assembly.

The rotating plate must be at an exact height above a reference plane, and thus it is also a precision piece of engineering. It should not be removed unless it is absolutely necessary to do so, since resetting it is very difficult. Generally, this is not necessary for head or tach-sensor changing.

The bearings used are of a very high caliber and are put in place by special processes. When a machine is turned off, the heads will continue to rotate for some time and should be allowed to do so. Damage to this part of the mounting usually means a replacement of the whole drum assembly.

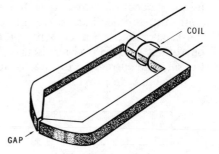

Fig. 6-12. Principle of video head.

The heads in a modern machine are of ferrite material and are shaped as shown in Figs. 6-12 and 6-13. The heads are about 10 mils thick and have a gap of approximately 40 to 50 microinches. Only about 6 to 8 turns of wire are used, and it is of microscopic thickness. This small ferrite chip is fastened with epoxy to a small metal plate, and this plate is used to attach the heads to the rotating plate (Fig. 6-14).

The head position is critical. Recording and playback cannot be started or relied upon until the head is positioned properly. Because of this, the mechanical mounting must be precise. But mechanical

Fig. 6-13. Typical video head.

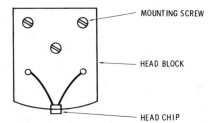

MOUNTING SCREW

HEAD BLOCK

HEAD CHIP

Fig. 6-14. Method of mounting
video head.

objects are not perfect, and some tolerances must always be permitted. In vtr's, the mechanical tolerances are very tight, and the minor inaccuracies are corrected by electronic pulse shaping and control in the head servo circuit. The radius through the head gap must make an exact angle with the magnet or vane that produces the head-tach pulse. Provision is made for adjusting this angle. To aid the servo in correcting for mechanical deficiencies, several electronic adjustments are provided.

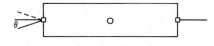

Fig. 6-15. Offset angle of video head.

θ = OFFSET ANGLE

Two-head machines must be provided with a further mechanical adjustment. The heads must be exactly 180° apart, with only about 3′ of tolerance allowed. This is shown as the offset angle (θ) in Fig. 6-15. An error of more than this will cause the two heads to begin scanning the tracks in slightly different places. The result is that the top of each successive field is displaced a small amount from the other. This effect is referred to as a *dihedral* problem. A screw adjustment is provided to correct this error, but usually the heads are preset so that it does not occur.

The two heads are mounted at each end of a bar, which is then attached to the rotating plate by two bolts. There are two mounting procedures in common use. In the first, the head bar is simply laid on the plate and bolted into place, and to correct for machining tolerances, a shim is placed under the head to achieve the correct height (Fig. 6-16). It is important to note that the shims go with the

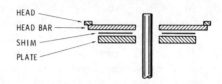

Fig. 6-16. Use of shim to adjust head height.

bar and not the plate. The second common method is to use the mounting screws to bend the bar to align the heads to the correct height. This requires a microscope height gauge. The tip projection in this type of head mounting is preset at the time of manufacture and usually is not adjustable.

HEAD DRIVE

The head disc or plate is mounted on a central shaft or spindle that is driven by a pulley on its lower end. This pulley is usually belt driven by a motor, but occasionally rim and direct have been used. With belt drive, the head pulley has a slightly smaller diameter than the motor pulley; thus the drive ratio is set to give a head free-running speed one percent faster than the correct recording speed. This allows a simple servo braking action to be used for speed control. It is important to note that the heads rotate in a direction opposite to the travel of the tape, and to achieve this the belt is often required to be twisted.

With direct or rim drive systems, the motor speed is directly controlled by the electronics, and separate braking is not used.

HEAD DEGAUSSING

With constant use, the heads become magnetized. This can cause noise in the picture and partial erasure of a previously recorded tape. An audio type of head demagnetizer can be used on the video heads in exactly the same way as it is used in an audio machine. However, do not touch the heads with the demagnetizer, and also check to ensure it is not too powerful. Touching a video head with a powerful head demagnetizer can actually shatter the ferrite chip. In color recording, it is extremely important to have thoroughly degaussed heads.

HEAD CONNECTIONS

Either slip rings or rotary transformers may be used to carry the signal to and from the heads. Slip rings are usually mounted on a post, and wire leaf springs make the electrical contacts, as in Fig. 6-17. Rotary transformers are mounted in the center of the head assembly (Fig. 6-18); one set of coils rotates with the head plate, and the other set is stationary. The windings can be around the vertical shaft or on a disc attached to the shaft.

Rotary transformers have the advantage of no mechanical noise, and they can be used in machines where head changing does not require the removal of the whole drum. In this way, they can be protected and shielded for optimum performance.

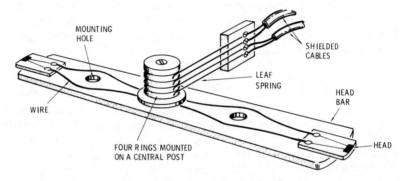

Fig. 6-17. Slip rings for video heads.

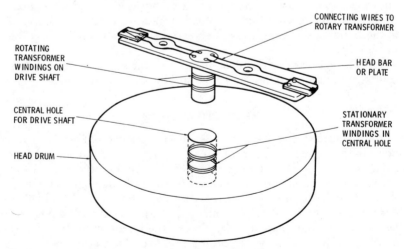

Fig. 6-18. Rotary transformer for video heads.

HEAD CHANGING

Since almost every machine has a different head arrangement, each one has a different procedure for changing the heads. These range from slipping the old head out and pushing the new one in, to complicated measuring and setups with hair-line microscopes to ensure correct mechanical dimensions. Replacement is required when the heads are damaged, worn out, have open coils, etc., or when sufficient tape penetration cannot be obtained.

Head changing is so different for the individual models that it is best left to the service manuals. On some machines, the head drum is never disassembled, but on others it is necessary to remove the top half to replace the heads. This removal and replacement requires much care. Correct tape tracking requires that the upper and lower halves of the head drum be absolutely parallel and concentric with the central drive shaft and the heads. This requires very precise machining and polishing of the drum surfaces and equally accurate placing of the two halves.

Since the top half is made to be removed for servicing, some locating indication must be used for its replacement. For this, a carefully positioned locating plate or butt at the rear of the assembly is used as a position reference. If this is not built into the machine, it is advisable to scribe the drum and the permanent part of the assembly to which it attaches before removal, as in Fig. 6-19. Two such scribe marks can ensure accurate replacement.

If the top half of the drum is not correctly replaced, the result is that the tape is held away from the drum at some points (Fig. 6-20A), and thus the head is prevented from scanning this part of the tape. The result is noise and no picture on part of the screen (Fig. 6-20B), and the same thing can be observed on a waveform

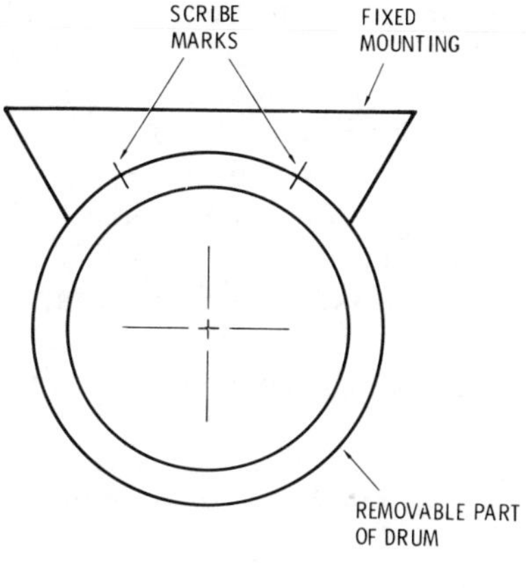

Fig. 6-19. Scribe marks for realignment of head drum.

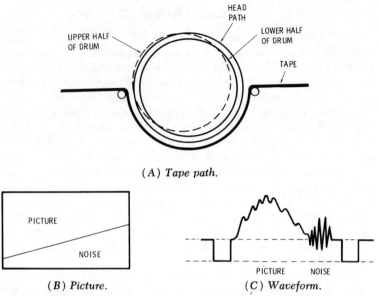

(A) *Tape path.*

(B) *Picture.*   (C) *Waveform.*

Fig. 6-20. Effect of incorrect replacement of top half of head drum.

monitor (Fig. 6-20C). Fortunately, it is possible to correct this problem by viewing a picture or waveform monitor while adjusting the drum.

Unless the manual specifies it, the head drum assembly should never be disassembled, and none of its components should be touched. Only the video heads should be removed, and this should be performed only in the approved manner, using any approved tools and alignment jigs that are required.

Once the new head is in place, penetration adjustment is necessary. Dihedral adjustment may also be required. Height is automatically set by the manufacturing dimension of the head and plate or by placing shims under the heads or head plate.

To perform head removal and replacement, and to make the necessary adjustments, a set of tools is often provided by the manufacturer. Fig. 6-21 shows a simple tool for pulling heads out from the side of a head drum. Another item is the head penetration gauge. This is a curved metal block shaped so that it can be held against the head drum but still allow the heads to be rotated slowly and without damage. Mounted on the block are a weak leaf spring, a sliding pin, and a meter. As the head moves past the block, it pushes the spring away, and this moves the pin and the dial needle. In this way, the distance the head penetrates is read directly on the meter dial.

94

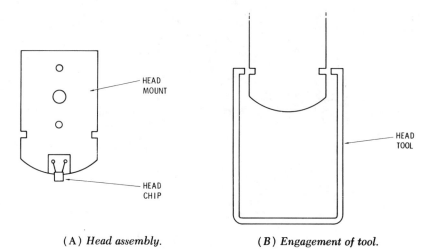

(A) *Head assembly.*   (B) *Engagement of tool.*

Fig. 6-21. Special tool for head removal.

First, the dial must be calibrated by holding it against a precisely machined block (Fig. 6-22). The two are held in contact, and the dial is adjusted to read zero. After calibration, the head penetration can be set. The calibration block is mounted on the upper part of the drum so that they fit flush together (Fig. 6-23). The plate is moved so that the head passes the leaf spring and deflects the dial needle. The maximum deflection is the head penetration beyond the drum circumference. It can be adjusted by using an eccentric driver (Fig. 6-24) inserted through the correct hole in the head. The head is adjusted to have a penetration of 100 ± 10 micrometers. Once this is set, the video head is tightened in position.

The two heads are inserted and adjusted in the same way. Once in position, they should be exactly 180° apart. This is checked by the

Fig. 6-22. Dial gauge calibration.

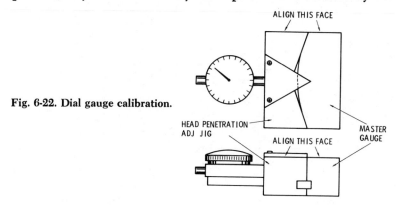

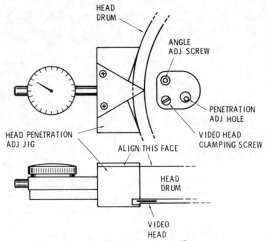

Fig. 6-23. Head penetration adjustment.

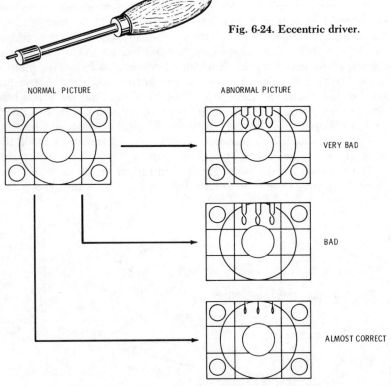

Fig. 6-24. Eccentric driver.

Fig. 6-25. Use of picture monitor to check head alignment.

simple means of viewing the picture on a tv screen. If the heads are not 180° apart, a flagging will be seen at the top of the picture, as in Fig. 6-25. The cure is to adjust one head only to make the picture perfect. This is a simple trial-and-error process in which tapered bolts are inserted or a special screw adjusted to twist the head about its own mounting screw.

THE CAPSTAN ASSEMBLY

The capstan assembly of a vtr is no different in principle from that found in an audio recorder, and it is often exactly the same in practice. Fig. 6-26 shows the simplest and most common arrangement. The pressure-roller arm is actuated by a cam-type arrangement, and the roller is held against the steel capstan by spring pressure. The capstan is placed after the other functions and is the last item before the take-up spool, so in effect it drags the tape past the heads and the guides. The flywheel is carefully balanced so that the capstan runs truly concentric around its central axis and thus provides an even drive without introducing wow and flutter. A belt from an electric motor is the most common form of drive. This same motor is often used simultaneously to belt drive the heads and to provide all the other tape-transport functions.

The roller is made of synthetic rubber and is bearing-mounted to ensure smooth running. Usually, it has an even matt surface to provide good contact with the tape and capstan. Some machines have a serrated roller (Fig. 6-27) to prevent an air bearing from forming between the tape and roller. Such an air bearing can introduce instability problems. The roller may operate against the backing or the

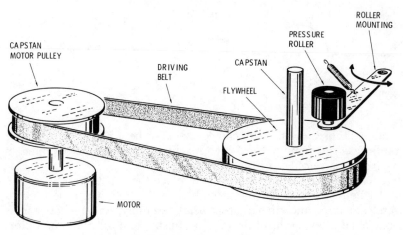

Fig. 6-26. Basic parts of capstan assembly.

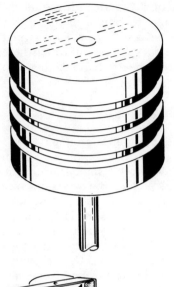

Fig. 6-27. Serrated pressure roller.

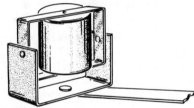

Fig. 6-28. Pressure roller in
pivoting assembly.

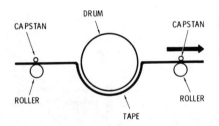

Fig. 6-29. Use of two capstans in vtr.

oxide side of the tape; the choice seems to be a matter of designer preference. In some of the 1-inch machines, the roller is mounted in a pivoting assembly as in Fig. 6-28.

A commonly used alternative is to place the capstan and pressure roller before the head-drum assembly. In this way, the tape is accurately metered into the rotating-head assembly, and the tension is controlled after the head drum. The capstan and roller are no different in this arrangement.

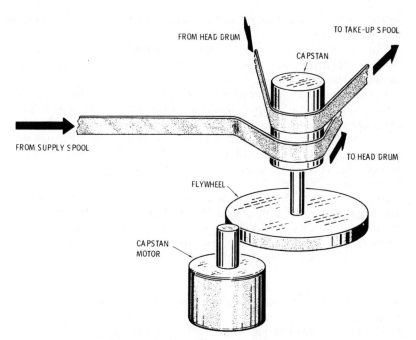

Fig. 6-30. Use of rubber-covered capstan.

A few machines have been produced with two capstans and rollers, one mounted before the drum and guides and one after as illustrated in Fig. 6-29. The capstans are separately driven to provide the correct tape speed and tension.

In Ampex machines, the conventional capstan and roller are dispensed with, and the tape is wrapped twice around a rubber-covered capstan of wider than normal diameter (Fig. 6-30). The tape is guided almost halfway around the lower part of the capstan before entering the head drum, and on leaving the head drum it wraps almost halfway around the upper part of the capstan. The oxide side of the tape is in contact with the capstan, and this large area of contact provides a positive drive. The capstan is rotated by a rim drive from a separate capstan motor.

Provided the capstan and roller are kept clean and not damaged, all the methods described will produce good results.

TAPE TENSION

Tape tension is one of the most important aspects of the mechanical and electrical design of a vtr. The tape tension must be correct on both record and playback, and on playback it must be adjustable so that it is the same as on record. This last factor is important in the interchange of tapes between machines.

The first sign of incorrect tension is hooking at the top of the picture. If the problem is not severe, adjustment of the control knob usually will cure it.

In the set-up and maintenance procedures for a vtr, there is a definite section on tape tension. The back tension on the tape as it is fed from the supply reel must be set within very narrow limits, as must the take-up tension. Excessive tension can stretch the tape and thus ruin it, as well as cause rapid head wear. Too little tension will cause loss of contact over the guides and scanning-head assembly, and thus prevent recording or playback from occurring.

Several factors affect tape tension. The tape changes its length and rigidity by stretching, and this varies considerably with temperature, humidity, and the applied tension. To further complicate matters, some machines use a constant tape tension, and others use a constant torque applied to the feed reel, which varies the tension as the tape diameter changes. This causes trouble when playing back on one type of machine a tape recorded on the other.

Tension can be applied to the tape by either mechanical or electrical means. Mechanical means are limited to the application of braking to the feed spool. The electrical means are usually a motor controlling the feed spool, a motor on the take-up spool, or the use of two capstans. The following paragraphs examine the more common methods in more detail.

Brake on the Feed Spool

A brake operating on the feed spool is used in most of the smaller and less expensive machines. This simple form of tension control is achieved by applying friction to the feed-spool table. The friction is applied in two ways. In one method, the spool table is in two parts with the top part riding on a felt pad that separates it from the stationary lower part. This applies a constant back drag to the tape. In the second method, tension variation is obtained from a brake band wrapped partially around the upper part of the spool table. The band is controlled by a pivoted arm (Fig. 6-31), on one end of which is a post that acts as a tape guide.

On record, an average setting is maintained throughout the length of the reel by having the arm remain in a central position. On playback, the same average setting is maintained, but it is varied slightly by the arm, which is now free to react to the tape. If the playback tracking is imperfect due to improper tension, it can be corrected to some extent by varying a "tension" or "skew" control. This is a knob that mechanically moves the neutral position of the arm, thus changing the effect of the braking band on the reel table.

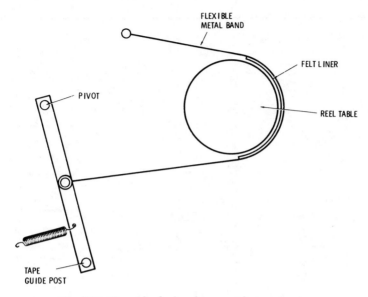

FLEXIBLE
METAL BAND

FELT LINER

PIVOT

REEL TABLE

TAPE
GUIDE POST

Fig. 6-31. Use of brake band to provide tape tension.

To ensure success with this method, the setup described in the service manual must be followed to the letter. If this does not produce the desired results, replacement of parts is often the only answer.

Typical values of tension in the tape should be 30 to 45 grams for a full reel and about 65 grams for an almost empty reel. This check is made with a spring-tension gauge in the normal maintenance alignment.

Take-Up Spool Control

Take-up spool control will be described by using the IVC 900 machine as an example. A feed capstan is used, and the tension is controlled after the head-drum assembly by using both a motor on the take-up spool and a motor-driven tension arm. Fig. 6-32 shows this arrangement.

In record, a constant tape tension is used, and in playback this is duplicated exactly. Three means of control are provided to achieve this:

1. In the record mode, a constant 8-ounce tension is used.
2. In the playback mode, an automatic control is active. It uses the video sync from the tape and the existing tension in the tape as inputs to a control servo.
3. A manual adjustment is provided.

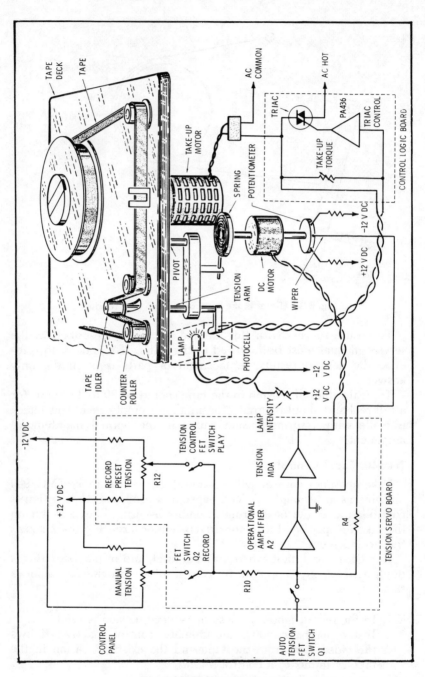

Fig. 6-32. IVC tension servo.

The tension in the tape is determined by the take-up torque and the force in the tension arm. The tension control consists of four elements: (1) the take-up reel motor, (2) the triac circuit that controls the motor, (3) the tension control assembly, and (4) the tension servo control.

The tension in the tape determines the position of the tension arm. Connected to the arm is a light-sensitive resistor separated from a fixed lamp by a distance that varies as the arm moves; hence the resistance is a function of the tape tension. This resistance is in the input circuit of an amplifier that controls the firing point of a triac, which in turn governs the ac power to the take-up motor. Hence, the take-up torque is directly dependent on the tape tension.

The pivoted tape-tension arm develops a force determined by a spring which is wound or unwound by a dc motor attached to its center. The drive to this motor is servo controlled, and the two inputs to the servo are a potentiometer on the motor and a reference signal derived from the vertical-sync on the tape. The arm thus moves to aid in keeping correct tension in the tape.

A control is provided for presetting the tension in the record mode, and a manual control is provided for playback.

Motor Control of Feed Spool

This example of motor control of the feed spool is taken from the Sony EV 210. A dc motor provides belt drive to the supply spool. The motor has a multigap head built into it, and this produces a sine-wave output as the motor rotates. This sine wave is amplified, limited, passed through a low-pass filter, and then further amplified (Fig. 6-33). It is then converted to a dc level by a detector circuit and used to control the drive to the dc motor. Changes in the tape tension are sensed by the motor and result in a varying frequency from the multigap head, which thus causes a variation in the dc from the detector; this is the feedback part of the servo loop. The reference is provided from two potentiometers. One is preset to provide the tension in the record mode, and the other is a manual control for the playback mode.

MOTORS

All the tape-transport and mechanical-drive functions in a helical vtr are provided by electric motors. The motors serve four main purposes:

1. To rotate the video heads
2. To rotate the capstan and roller
3. To rotate the tape reels
4. To provide servo controls where needed

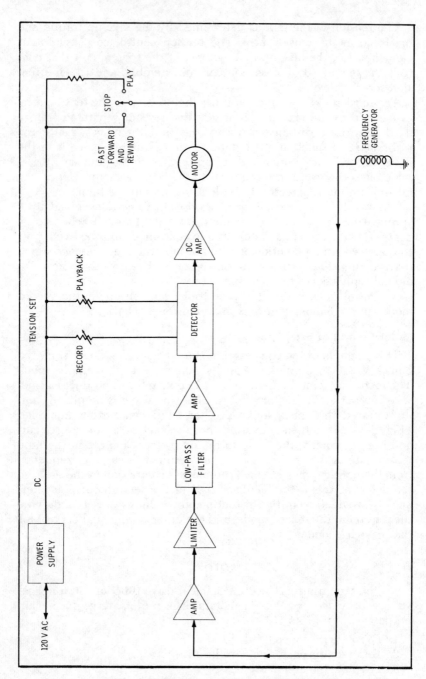

Fig. 6-33. Tension servo in Sony EV 210.

104

Most machines do not use a separate motor for each purpose; often only one is used to serve all the functions required. Two motors are commonly found, but three or four are somewhat rare.

The motors can be either ac or dc, with the ac motors being either induction or synchronous. The purpose for which the motor is used governs the choice of type. The choice of a motor and the control circuit is made from a consideration of the electronics of the servo system. It does not matter mechanically which is chosen, since the mounting and mechanical drives remain the same.

The Head-Drum Motor

A dc motor or an ac synchronous motor can be used to drive the head drum. The important thing is the use of a motor the speed of which can be accurately controlled. The motor is usually mounted separately from the head drum, and motion is provided through a belt drive. A hum belt is often used on these motors to minimize the effects of vibration on the entire system.

The Capstan Motor

In the simplest machines, an ac synchronous motor powered by the 60-Hz line is used to belt drive both the heads and the capstan. However, if the machine is to have an editing capability, a separate capstan motor is essential.

When a separate capstan motor is used, it is invariably driven from a servo circuit. The choice is usually limited to a dc motor or an ac synchonous motor, since these are the easiest to be speed controlled.

Tape Reels

The larger, more expensive machines often have a separate motor for reel drive and other mechanical functions that are not part of the head and capstan motion. Smaller machines use the capstan or head motor for the fast-forward and rewind modes. On record and playback, the tape reels are seldom driven by a separate motor; the takeup reel is belt or rim driven, and the supply reel is caused to rotate by the action of the tape being pulled from it.

Servo Control

Head and capstan servo control is most often applied directly to the motor or to a braking system. Occasionally, the speed control is provided by a belt drive from a small subsidiary motor.

BELT DRIVE

The use of a rubber or plastic belt between a motor shaft and a drive wheel is common throughout the tape-recorder industry. It provides an excellent method of driving a rotating mechanism at a constant speed. It is a good enough method of drive to be used in vtr's and to provide the necessary very fine control required for the head and capstan servos.

In most modern machines, a belt drive from the capstan motor to the flywheel is used. This ensures a constant speed, and when a momentary change of speed is required, it permits the change to be made without overshoot and instability.

Even though the belt drive has been around for a long time, it offers a number of advantages:

1. A large amount of power can be transmitted in this way.
2. If part of a machine stalls, the belt will slip, thus preventing damage to a whole mechanism.
3. It is capable of very stable drive.
4. If speed control is required, it can be achieved by utilizing a controlled amount of slippage between the belt and metal.

Belts are usually polished on one side only, the other side being left rough. The coefficient of friction between the belt and metal surface is very important, and the belt must be installed with the correct surface in contact with the metal. In most cases, this is the rough side, but some exceptions do exist. If the wrong side is used, the slippage will be wrong, and the tape transport will not operate correctly. When changing a belt, carefully check the service manual and the belt being removed.

An old belt will often be longer than a new one, because continual use causes some stretching. The easiest way to check this is to hang the belts next to each other over a pencil or finger. When installing a new belt, be careful so that it is not stretched. Also, all belts and surfaces should be clean and free from dirt and grease.

RIM DRIVE

Rim drive is not commonly found in the vtr for head and capstan drive, but it is used extensively in driving the reel tables. It is usually found in the form of a rubber idler or "puck," interspersed between two other surfaces, such as metal drive spindles and plastic wheels on a reel table. Rim drive is suitable for this type of application since it can give a good, firm drive and can be easily disengaged for change of function.

REEL TABLES

The reel tables in vtr's do not differ greatly from those found in audio machines. Belt, puck, and separate motor drives are used along with friction braking by means of brake stops, bands, and fiber washers. Adjustments are provided to obtain correct tension from both reels in all modes of operation. Disassembly is relatively simple with few component parts, and little maintenance is required.

MECHANICAL LINKAGES AND LEVERS

The various functions and modes of operation are selected either by a lever or by solenoids controlled by push buttons. Whichever method is used will require a system of internal mechanical links to effect operation. These take the form of rods, levers, cams, pucks, belts, etc., which engage each other in a series of moves to connect motors, tables, capstans, etc., and to actuate switches to provide power and signal paths. Almost every model uses a different arrangement, and these vary from simple to extremely complex.

The electronic portions of the machine are usually mounted on printed circuit boards, with the different circuits on different boards or parts of boards. These are switched in and out as required by switches mounted among the mechanical levers, or by slide switches on the printed boards. Relays are sometimes used for signal switching, and these are powered by switches in the mechanics.

THE OTHER HEADS

In addition to the scanning video heads, several other heads are required in a vtr. These are:

1. Audio heads for record, playback, and erase
2. Control heads for record, playback, and erase
3. The main erase head

The audio heads are similar to those in an audio tape machine. A combined record/playback head gap is mounted in a housing slightly wider than the tape. The gap is placed to be correctly aligned with the audio track on the tape. The audio erase head gap is mounted in front of the record/playback gap, either in the same housing or a separate one. The electronics of audio record, playback, and erase are exactly as in an audio recorder.

The control heads are exactly the same as the audio heads. They are often mounted in the same block as the audio heads, but the gap is placed so that it scans the opposite edge of the tape. A control erase head is not always provided, since it is needed for insert editing only.

The main erase head is placed on the other side of the head drum from the audio and control heads. Its gap is large enough to span all or most of the tape, and its function is to clean the tape of pre-

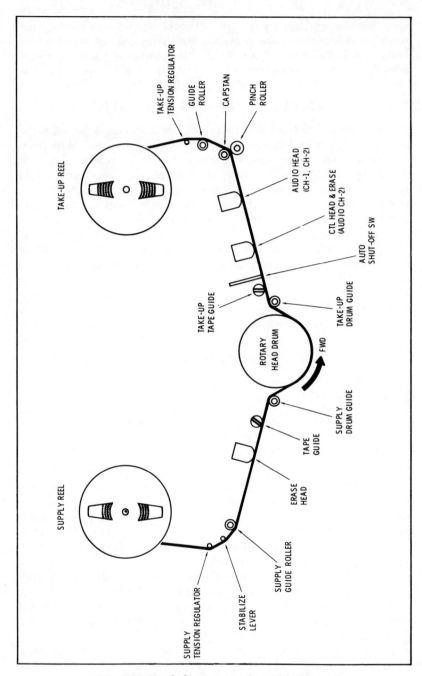

Fig. 6-34. Head placement on Sony EV-320 vtr.

recorded material prior to passage of the tape over the video heads. In a simple nonediting machine, this head spans the whole tape, but in an editing machine it does not cover the audio tracks, and not always the control tracks. The main erase head is supplied either with the same erase signal as the audio and control heads or with an rf bias from a separate oscillator.

The lateral position of the audio head with respect to the head drum is important. If the audio head is too far off position, lip sync can be lost. The mounting of the audio head is such that this usually cannot occur, but provision is made for both azimuth and upright adjustment. When a head is completely changed, the height of the new head can be set by screws or shims so that the gaps align with the tape tracks.

The lateral position of the control head is of critical importance. If this head is too far from its correct position, the servo cannot lock up because the time relation of the head-tach pulses to the control pulses is wrong. The control head can be moved sideways, and this should be checked whenever a replacement is made. The easy way to do this is to measure the rf level at the input to the first fm limiter. The head is moved sideways until this is at a maximum.

The choice of which heads to have on a machine and of their physical arrangement is dependent upon the use for which the vtr is intended. There is nothing standard in the number of heads used, their placing, construction, etc. The only requirement is that they be the same on any particular model. The general placement is as shown in Fig. 6-34, but for a specific model, the service manual should be consulted.

7

The Record and
Playback System

This chapter covers the electronics of the record and playback process. The other electronic functions, such as servos, are covered in other chapters.

The two main functions of a vtr are the recording and playback of a video signal. The input to the machine must be the standard video signal common to the tv industry, and whatever method is adopted for storing this signal, it must be converted back to a standard signal on playback.

When the tape is wrapped in a helix around the head drum, the record head scans slanted tracks across the tape, and each track contains one complete field of the tv signal. The tracks begin near the top or bottom of the tape with some point close to the vertical interval. The heads must be in an exact position at the correct time, and this is accomplished by the servo circuitry. The position of the head is in no way governed by the recording amplifiers, and these will continue to supply current to the heads and record a signal on tape even if the servo is entirely malfunctioning. The control of dropout periods, changeover times, and other functions is also part of the servo system and has nothing to do with the record or playback circuits.

This chapter will first examine the video record process and will follow this with a description of the playback methods.

VIDEO RECORDING

Fig. 7-1 is a simplified block diagram of the basic recording chain. The standard video signal is applied to the first stage of the video

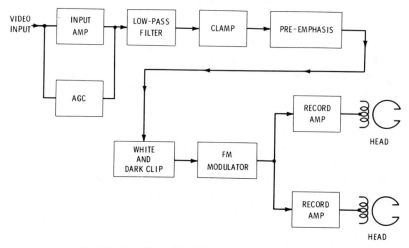

Fig. 7-1. Simplified block diagram of recording chain.

record amplifier, where it is amplified from its 1-volt peak-to-peak level to some specified larger level. This first stage passes all the information contained in the signal, including the 3.58-MHz color information and even the 4.5-MHz audio carrier if it is present.

The signal then passes through a low-pass filter that has a cutoff point around 3.5 MHz. This removes the chroma and audio. It is necessary to remove these components because they can cause interference patterns in the final picture. The signal is now further amplified. A voltage for automatic gain control (agc) is taken from this point or from the output of the first stage, and it is applied to a control amplifier that affects the gain of the stage. In this manner, a correct design level of signal is applied to all stages in the chain.

Until this point, the signal has been capacitively coupled, and for reasons which will be apparent later, it must be dc restored, or clamped. This is accomplished by clamping the sync tips to a given dc level by a sync clamp control.

The next stage is a pre-emphasis circuit. This stage boosts the high frequencies of the signal more than the low frequencies, because it is in the high range that noise becomes apparent in the electronic circuitry. Boosting at this point makes it possible for a better signal-to-noise ratio to be obtained. The pre-emphasis will not remove noise already in the signal, but it will eliminate further noise introduced by the vtr electronics and will obviate any tape noise. The pre-emphasis circuit produces overshoots in the signal at those places where the signal changes most rapidly, so spikes appear on peak whites and at the sync tips. In order that these do not cause trouble, they are clipped off by a white clipper and a dark clipper. After clipping, the result is a video signal with accentuated high frequencies and a definite upper and lower dc level with respect to ground. It is ready to be applied to the fm modulator and put on tape.

Modulation can be accomplished in several ways, but the most common are to vary the frequency of a sine-wave oscillator or vary the frequency of a square-wave oscillator. The variation from the unmodulated frequency is directly proportional to the signal level, and the rate of variation is governed by the rate at which the signal changes level. The amount of frequency variation is known as the *deviation*, and it does not have to have the same frequency range as that of the signal. A 3-MHz bandwidth of input signal may cause a deviation of only 1 MHz; however, the *rate* at which the signal changes demands that the associated circuitry still have the capability to pass all of that signal. In other words, the bandwidth of the associated circuits must still be 3 MHz, even if less than this is used as a deviation.

The signal-to-noise ratio is improved by a large deviation, so the deviation is not severely restricted; but, since the deviation is governed by the bandwidth of the tape and the heads, some limitation is required. If the modulation is excessive, the white level of the input video will actually appear black on the reproduced image. And since white compression is much less noticeable and less objectionable than overmodulation noise, a reasonable restriction is placed on the frequency swing of the fm modulator.

With no signal applied to the modulator, it is critically important that the output should be as perfect a sine or square wave as possible. If it is unbalanced, a beat will be created on playback and E-E; this will be seen on the screen as a moiré pattern. Since the demodulation process at one point doubles the frequency of the carrier to help in recovery of the video signal, and unbalance becomes very prominent, it is again important that the signal should be symmetrical. To achieve this, a carrier balance control is provided.

From the modulator, the signal passes to the record amplifier. The input of this amplifier has a level control that regulates the current applied to the video heads. The record current has an optimum level, and the reproduced voltage decreases if the record current is either above or below this level, so on setting up a machine this becomes an important adjustment.

The signal is finally applied to the heads through a slip-ring assembly or through rotary transformers. Both methods give excellent results, but the slip rings can be the cause of some noise due to dirt or bad contacts. However, they are much easier to replace and clean than are defective rotary transformers.

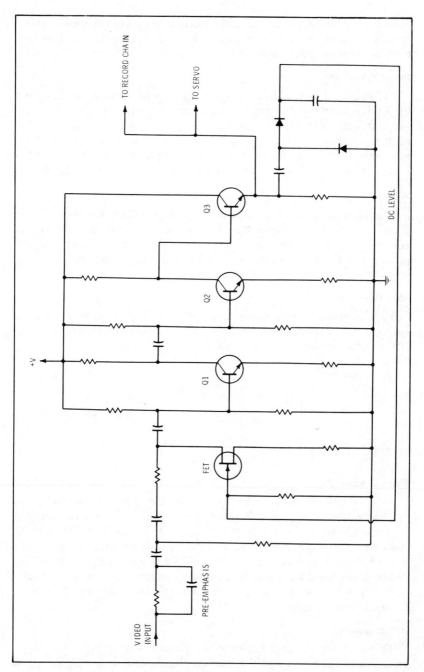

Fig. 7-2. Circuit of input amplifier.

In the following paragraphs, the various stages of the video record circuit are analyzed in more detail. For clarity, simplified drawings from several machines are used.

Input Amplifier

In the circuit of Fig. 7-2, the video input is pre-emphasized slightly by an RC network before it is applied to transistor Q1. This pre-emphasis is provided so that the agc will work adequately on the high-frequency portions of the input. Otherwise, this is an ordinary video amplifier. The output is fed to three places: the low-pass filter in the recording chain, the servo sync stripper, and the agc circuit.

AGC

The agc, or automatic gain control, is provided by the FET in Fig. 7-2. Part of the output signal is rectified to obtain a dc level that controls the FET. The FET acts as a shunt and drains varying amounts of the input signal to ground. This system maintains a constant level of output from the amplifier.

Low-Pass Filter

The low-pass filter is needed to reject the high-frequency part of the signal, which contains the 3.58-MHz chroma carrier and the 4.5-MHz audio carrier. The circuit is shown in Fig. 7-3.

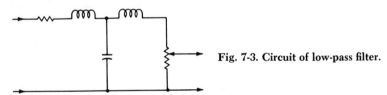

Fig. 7-3. Circuit of low-pass filter.

Sync Clamp

After it goes through the low-pass filter, the signal is isolated by an emitter follower. Up to this point, the signal has been ac-coupled, and it is necessary to restore the dc level. Assume the video at point A in Fig. 7-4 is positive going and the sync is negative going. Then the most negative point of the sync will be determined by the setting of the potentiometer. From this point on to the input of the modulator, only dc coupling is used, and reference levels are needed to set the frequency swing.

Pre-Emphasis

To obviate noise and improve the signal-to-noise ratio, the high-frequency end of the signal is pre-emphasized. This is accomplished in the simple circuit of Fig. 7-5.

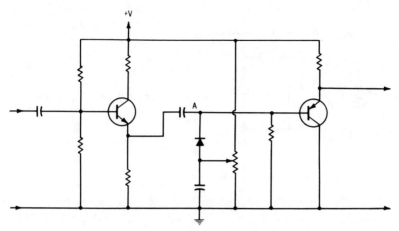

Fig. 7-4. Circuit of sync clamp.

Fig. 7-5. Pre-emphasis circuit.

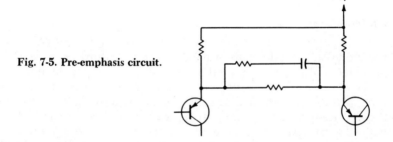

White and Dark Clip

The pre-emphasized video contains overshoots in both the positive-going and negative-going directions. These occur as spikes and must be removed in order not to distort the picture. The spikes are apparent in the waveform of Fig. 7-6. Fig. 7-7 shows a simple clipping circuit to remove the spikes that exceed a certain amplitude.

In some models, pre-emphasis occurs before the sync clamp. If this is done, the sync clamp effectively performs the function of the dark clipper, thus removing one control from the machine.

Fig. 7-6. Video waveform containing overshoots.

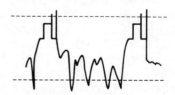

115

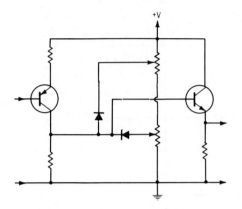

Fig. 7-7. Clipping circuit.

A white clipper must always be used. Although it produces white compression by removal of part of the signal, this is much less objectionable to the eye than the problems of overmodulation.

After clipping, the signal is ready to be fed into the modulator.

FM Modulator

There are two basic types of modulators used in helical vtr's: sine-wave oscillators and square-wave oscillators. A third type using the pulse-interval modulation method is found in IVC machines.

Whatever method is adopted, certain things are common to all types. All can be classified as voltage-controlled oscillators, in which the deviation is directly proportional to the level of the applied signal. All must have extreme accuracy in the symmetry of their unmodulated waveform. And, all must be absolutely frequency stable in their unmodulated state. Lack of symmetry produces moiré patterns on the screen. Frequency instability produces a varying level of illumination that can be observed on the screen with no video input and that still rides over the picture when a video signal is applied.

Sine-Wave Oscillators—At first sight, it would seem that a varying-frequency sine wave would be a suitable signal to record onto the tape, and this in fact is a good way to record the signal. However, a difficulty arises in trying to vary the frequency of a sine-wave oscillator. Sine waves are usually produced by a fixed-frequency resonant or feedback circuit that is quite stable in its operation. Methods do exist for changing the frequency from the natural value, but the range achieved is very small when reliability is required. A good example of this is in fm radio. An fm carrier is typically in the vhf band, near 100 MHz for example. The total frequency swing allowed for fm broadcasting is ±75 kHz from the unmodulated frequency, which is about ±0.075 percent of the carrier frequency. A video signal, which requires about 3 MHz of frequency shift, would represent only a ±1.5 percent swing of a 100-MHz carrier, which is still within reason, but the 100-MHz carrier could never be impressed onto the tape. For these reasons, direct modulation of a sine-wave oscillator is not used in video recording.

The method adopted is to use two high-frequency sine-wave oscillators operating at somewhat different frequencies. The outputs of these two oscillators are heterodyned together, and the difference frequency is extracted. In practice, two variations of this idea are used. In one, the oscillators are operated about 4 MHz apart in frequency, and only one of them is modulated. In the other method, the video signal is applied to both oscillators so that one increases in frequency while the other decreases.

The advent of certain semiconductors has made this an easy method to implement. The video signal is applied to a varactor in the tank circuit of the oscillator. (A varactor is a semiconductor diode the capacitance of which depends on the voltage across the diode.) Applying the video signal changes the capacitance of the varactor, and thus the frequency of the oscillator. Varactors work well at frequencies in the vhf range, making this the logical choice for the frequency of the oscillator and thus dictating the use of heterodyne methods to produce a signal that can be put onto tape.

The outputs of the oscillators are combined in a mixing circuit and then fed to a filter that passes only the small range around the difference frequency of the two oscillators. Thus the output of the low-pass filter is a frequency suitable for application to a recording-head amplifier.

A numerical example should clarify this procedure. Consider oscillators operating at 50 and 54 MHz. Then the difference frequency is 4 MHz, and the sum frequency is 104 MHz. Hence, it is easy to separate the 4-MHz output from the other frequencies. If the 50-MHz signal is varied to 51 MHz, the difference becomes 3 MHz; if it is lowered to 49 MHz, a 5-MHz signal is produced. So, whatever variation is introduced into the higher frequency is exactly repeated in the lower frequency. In this way, the original information is preserved at the lower frequency.

Both high frequencies can be varied, usually in the opposite direction. This means that each will be varied through a smaller frequency range than an individual oscillator, and thus each can work over a more restricted and linear part of its characteristics.

A typical unmodulated oscillator frequency is 50 MHz. After heterodyning down, a no-input signal will be around 2.6 MHz, the sync tip will produce 2.5 MHz, and peak white will range up to 5 MHz.

A typical circuit is shown in Fig. 7-8. This is a type in which both oscillators are modulated. A circuit with a single modulated oscilla-

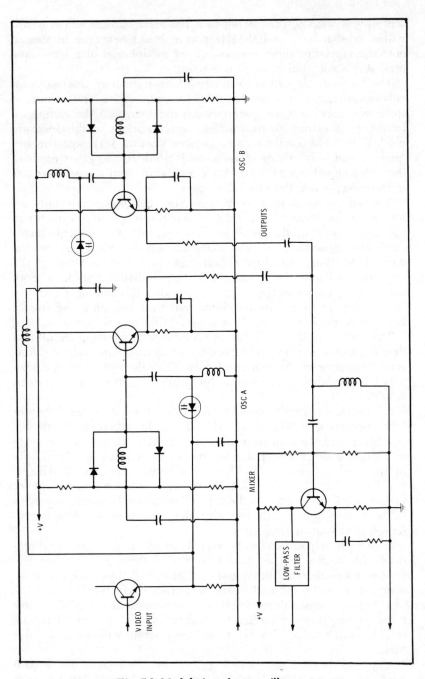

Fig. 7-8. Modulation of two oscillators.

118

tor is very similar, with the video signal being fed to one oscillator only.

Whichever method is used, the two outputs are combined and passed through a low-pass filter similar to that in Fig. 7-9. This will pass only the low difference frequency and effectively block all the higher frequencies. From the output of this filter, the signal is further amplified and fed to the record amplifiers and then to the heads.

Square-Wave Oscillators—When the sine wave of the preceding paragraphs is amplified and impressed onto the tape, it is amplified to the point where it saturates the tape, In this way, amplitude variations caused by the tape or the electronics will be obviated, and thus an important source of noise is eliminated. But now the signal on the tape is in the form of a sine wave with severe clipping of its peaks. In fact, the waveform is more like a square wave with not very steep sides. Since this is the condition of the signal on the tape, it makes some sense to use a square wave initially.

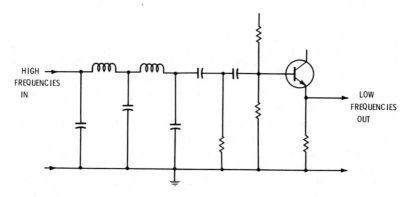

Fig. 7-9. Low-pass filter.

If a square wave is used, certain other advantages become apparent. It is just as easy to produce a square wave as a sine wave, and it is considerably easier to vary its frequency. Furthermore, the frequency variations are extremely linear over a wide range with respect to the signal that produces the variations. Waveform symmetry is easy to achieve and to adjust. But most important is that the un-modulated frequency can be produced directly without any form of secondary oscillator or heterodyning techniques.

The simple multivibrator shown in Fig. 7-10 will run at a frequency determined by the capacitors and resistors that cross couple the bases and collectors. Varying one of these components in either stage will vary both the frequency and the symmetry of the output. The easiest way to vary the frequency while preserving the sym-

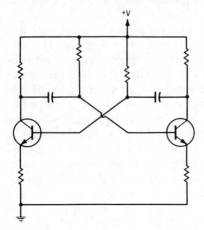

Fig. 7-10. Basic multivibrator circuit.

metry is to return the two base resistors to a separate voltage supply (Fig. 7-11) and to vary this supply.

If the control voltage is a video signal, the frequency variations of the multivibrator will be dependent on the video signal and will be an accurate measure of the video signal. When this multivibrator output is further amplified and processed, no precautions are needed to preserve its waveshape, and it can be used to saturate a tape without any form of signal distortion.

This simple method is now almost universally used in vtr's. It uses only small, inexpensive, reliable components, and in fact an IC can be used with a minimum of external components. Overall, it is easier and cheaper to implement than the sine-wave method and has several advantages. Fig. 7-12 is a typical practical circuit.

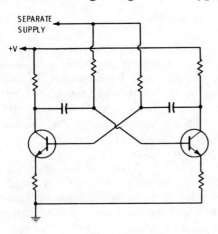

Fig. 7-11. Method of varying multivibrator frequency.

PIM Modulator—At first sight, the result of pulse-interval modulation looks like a normal frequency-modulated square wave; however, it is different in that it is the period and not the frequency that changes linearly with respect to the video signal. In pim, the duration of each semiperiod, or half cycle, is determined by the instantaneous amplitude of the modulating signal. Thus the interval is inversely proportional to the signal amplitude.

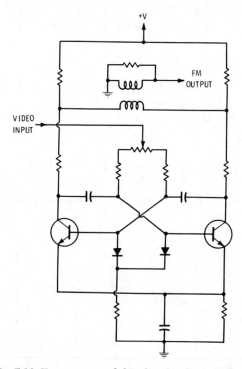

Fig. 7-12. Frequency-modulated multivibrator circuit.

The pim signal is produced in a circuit that looks like a modified differential amplifier but in fact is not. The pim signal is obtained by adding a constant-slope ramp to the incoming video signal. When the sum of these reaches a predetermined level, the output of the modulator changes state. The ramp is then reset to zero so that it can begin again. This produces a waveform such as the output shown in Fig. 7-13. This waveform looks like an fm square wave, but the information is encoded quite differently.

In order for the system not to be limited by the retrace time needed for the ramp to reset, two ramps are used. In this way, a fast sampling of the video signal is achieved.

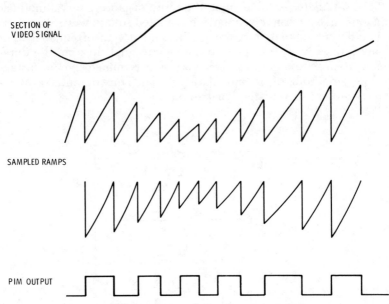

SECTION OF
VIDEO SIGNAL

SAMPLED RAMPS

PIM OUTPUT

Fig. 7-13. Principle of pulse-interval modulation.

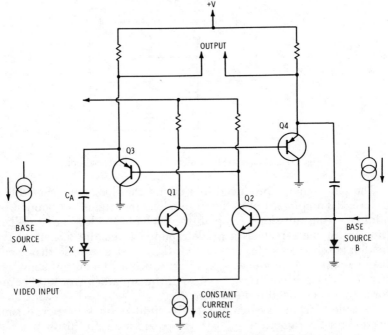

Fig. 7-14. Basic circuit for pulse-interval modulation.

Fig. 7-14 is a basic schematic of the pim modulator. Assume that Q1 is off and Q2 is conducting. Capacitor C_A is assumed charged to a predetermined voltage level, and is now discharging linearly due to the current from base source A. The video signal is applied to the common emitters, and Q2 acts as a linear common-base video amplifier with the video signal appearing at its collector. The video signal also appears at the top of capacitor C_A because of the emitter-follower action of Q3. So the voltage at the base of Q1 is the sum of the video signal and the constant-slope ramp of the discharge of capacitor C_A. Eventually, this reaches a voltage that causes Q1 to turn on and thus causes Q2 to turn off.

This action causes the state of the output of the modulator, as seen at the collectors, to change. This pulls the top plate of capacitor C_A up to a preset level, and since the bottom plate is clamped by diode X, the capacitor charges very quickly. Thus the capacitor is ready for use again long before the modulator flips back to its initial state.

Because the video is varying when added to the ramps, the sums will reach the given threshold at times dependent on the amplitude of the video signal (Fig. 7-13). This threshold is the trigger level for the modulator to change state, and so the output changeover times are governed by the incoming video. After the changeover, the action is the same, but with Q2 off and Q1 conducting.

The design is such that no transistors enter the collector saturation region, and thus the circuit is very fast in its action. The sync tips correspond to a frequency of 5.4 MHz, and peak white corresponds to 6.6 MHz.

Since the video signal appears at the collectors of the two transistors, it also appears in the output. It is removed by a nonsaturating balanced differential amplifier. The balanced signals at the two collectors of this amplifier are unbalanced by a transformer and then fed to the head record amplifiers. Fig. 7-15 is a simplified circuit of this amplifier.

Record Amplifier

The purpose of the record amplifier is to take the output of the fm modulator and apply it to the record head or heads. The method is the same in all machines, with differences of detail and not principle.

Figs. 7-16 and 7-17 are from one-head machines. The output of the modulator is set at the input of the amplifier by the record level control, which governs the amount of current fed to the head. The level is optimized during an alignment or maintenance procedure and then left at this setting. The optimum level is that which produces the maximum output from the head on playback. The heads are coupled to the amplifier output by a rotary transformer or slip rings.

123

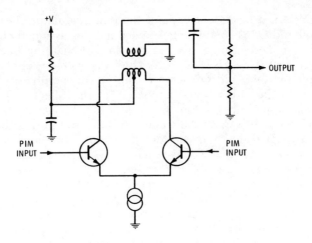

Fig. 7-15. Balanced differential amplifier.

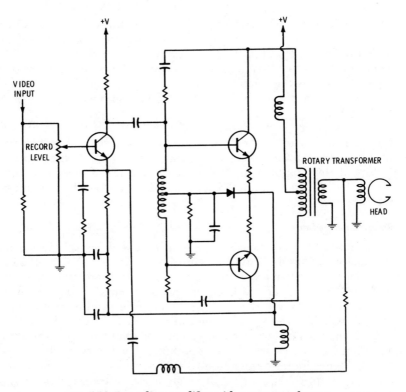

Fig. 7-16. Recording amplifier with rotary transformer.

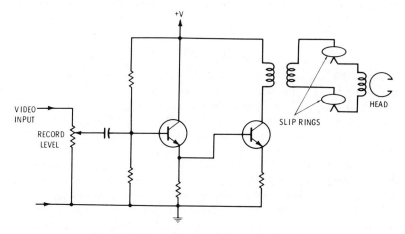

Fig. 7-17. Recording amplifier with slip rings.

A two-head machine is basically the same, but a minor difference exists at the output because of the two heads. Both heads can be fed from a single amplifier as in Fig. 7-18, with a balance control to ensure approximately equal currents to the heads. The record current and the balance control are both set for optimum playback during the maintenance alignment. Alternatively, the two heads can be fed

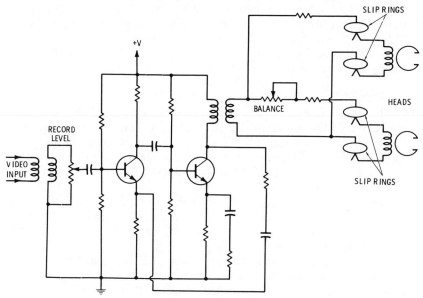

Fig. 7-18. Recording amplifier feeding two heads.

from separate amplifiers, in which case each head is individually optimized.

VIDEO PLAYBACK

Fig. 7-19 is a block diagram of the basic playback process. In the playback mode, the heads follow the tape tracks, pick up the recorded signal, and apply it through transformers or slip rings to the inputs of the playback preamplifiers. This is a very critical stage in the playback process, as it is here that the final quality of the image and its signal-to-noise ratio are determined.

In a one-head machine, the output of the preamplifier is applied directly to an fm limiter. In a two-head machine, the outputs are alternately switched and balanced before application to the limiter.

The input to the limiter is a very weak signal. The limiter performs two functions which at first seem to be in conflict. Each stage amplifies the signal, and then it limits the size of this output to a maximum by chopping off the signal at a given level. Several cascaded stages are used, with each performing the same function, so that whatever signal is fed into the limiter, its output will be a clipped sine wave or square wave with a definite peak-to-peak amplitude. It will not have any amplitude variations, thus removing an important source of noise and interference. The only information contained will be the frequency variations of the fm signal.

The signal out of the limiter is demodulated and converted back to a video signal. The video is then de-emphasized and applied to a video amplifier, which usually has several outputs.

The following subsections cover the individual parts of the playback chain in more detail.

Preamplifiers

The output from the video heads is a low-level fm signal, which is connected to the inputs of the preamplifiers. This low-level signal is easily affected by noise and extraneous signals, so considerable care must be taken in coupling it to the preamplifiers. For this reason, the preamplifiers are often mounted in the top of the head drum assembly, directly over the heads.

The preamplifiers have high-impedance inputs and high gain. Often, FETs arranged in a cascode configuration are used. Two or three subsequent stages are usually sufficient to provide an output suitable to drive the first limiter stage.

Three common arrangements exist for the preamplifiers in vtr's: A one-head machine needs only one preamplifier; a two-head machine may have two unswitched preamplifiers, or it may have switched preamplifiers.

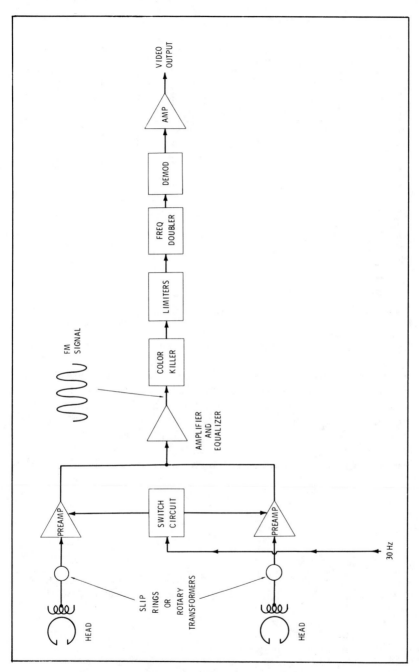

Fig. 7-19. Simplified block diagram of playback chain.

127

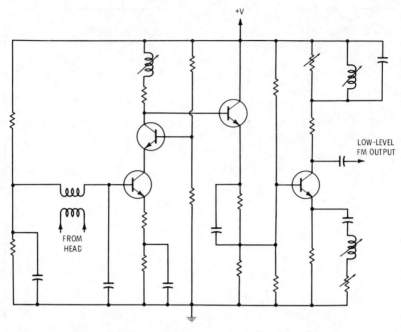

Fig. 7-20. Preamplifier circuit in one-head machine.

One-Head Machine—Since one head is scanning the tape continuously, only one preamplifier is required. Fig. 7-20 shows a typical circuit. Bipolar transistors are used in cascode to achieve a high signal-to-noise ratio, and feedback is used to reduce the input impedance. This is done to reduce the Q of the head and thus provide a more uniform response that is independent of the head characteristics. This obviates problems which could arise when the head is

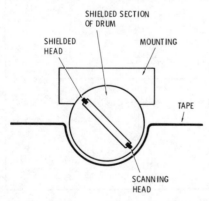

Fig. 7-21. Principle of two-head machine without head switching.

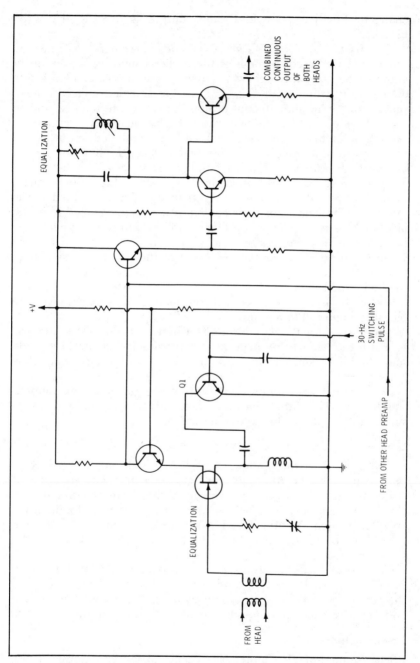

Fig. 7-22. Preamplifier used in machine with head switching.

changed. Finally, a low-impedance output is used to drive the limiters.

Two-Head Machine With Unswitched Preamplifiers—This arrangement was used in some of the earlier machines. As one head scanned the tape, the other was passing around the shielded inside of the head drum, as in Fig. 7-21. In the shielded area, the head picked up a minimum amount of noise or other interference, and so the output of its preamplifier was virtually silent. The outputs of both preamplifiers were paralleled, and the combined output signal was applied to an amplifier before being fed to the limiters.

Two-Head Machine With Preamplifier Switching—The head coupling and first stages are similar to those used in the other arrangements, but a switching transistor (Q1 in Fig. 7-22) is added at the first stage. This transistor is controlled by a pulse developed from the head-tach outputs. The preamplifier is allowed to conduct while the head is scanning the tape, but it is shut off for the period of time when the head traverses the back part of the drum. The switching time and position in the picture signal are covered elsewhere. This method produces a fast, positive switch from one amplifier to the other, with almost no noise. The outputs of the preamplifiers are paralleled and fed to the limiters.

The purpose of the variable resistor and capacitor at the input is to cause the head to resonate at a particular frequency to provide some high-frequency compensation for the head-to-tape characteristics.

All preamplifiers contain some form of rf equalization to compensate for the head output characteristics. This is achieved by methods similar to that shown in the circuit of Fig. 7-22.

Limiters

The purpose of the limiters is to eliminate all amplitude fluctuations caused by head-to-tape contact and extraneous signals and noise. The input is a low-level fm signal that is amplified and then clipped so that its excursions are limited by diode pairs to about 0.7 volt. The simple circuit of Fig. 7-23 accomplishes this. The first stage is followed by a string of three or four similar stages, each with the same function.

The last stage allows a larger signal, about 3 volts peak to peak, to be formed at the output, and this is fed to the demodulator. In the final stage, a balance or symmetry control sets both the positive and negative excursions equally about a reference level or ground.

Demodulator

The input from the limiter is of sufficient level to allow the demodulator to produce a noise-free output that is a standard video

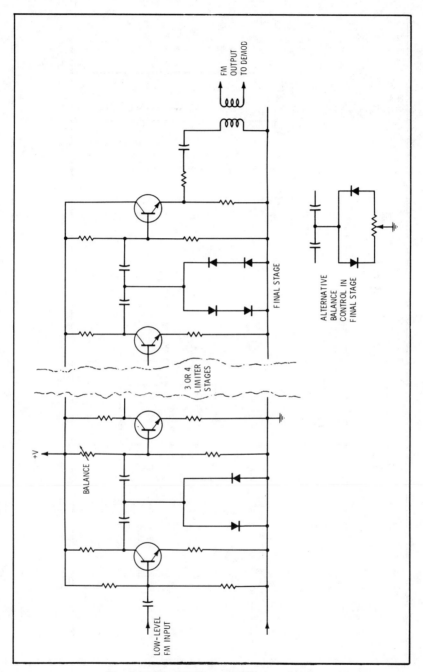

Fig. 7-23. Partial diagram of limiter circuit.

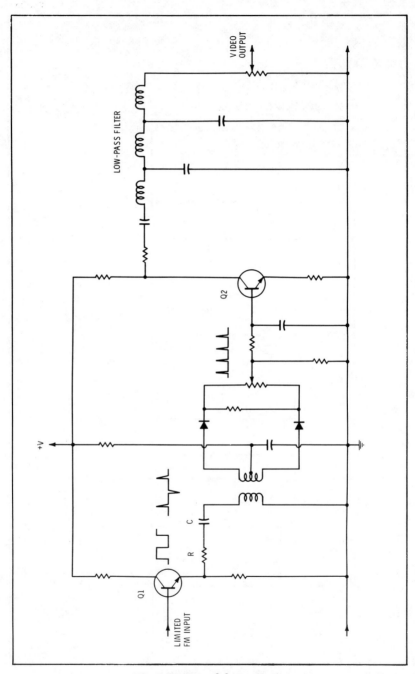

Fig. 7-24. Demodulator circuit.

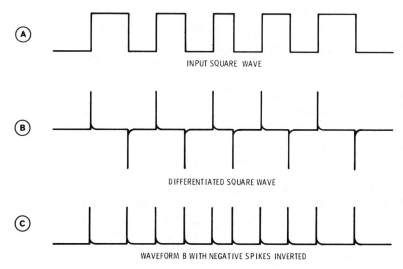

(A) INPUT SQUARE WAVE

(B) DIFFERENTIATED SQUARE WAVE

(C) WAVEFORM B WITH NEGATIVE SPIKES INVERTED

Fig. 7-25. Waveforms in demodulator.

signal. Most vtr's use the same method of accomplishing this end.

Reference to Fig. 7-24 will help in following this explanation. The varying-frequency square wave (Fig. 7-25A) at the input is differentiated by resistor R and capacitor C to produce sharp spikes (Fig. 7-25B). These are both positive going and negative going. The transformer and diodes act to invert the negative-going spikes. The potentiometer is set so that the inverted and noninverted spikes are picked off with equal amplitude, and they are then fed to the input of Q2. At this point, the spikes are all positive going and occur at twice the frequency of the input square wave (Fig. 7-25C). The video will be recovered from this train of spikes, so the carrier frequency from which the video is recovered has been placed outside the bandwidth of the video frequencies. By balancing the size of the spikes in an unmodulated waveform, all interference patterns can be removed from the screen. Without this rather odd technique, herringbone patterns can be seen on the screen, and they are very difficult to eliminate.

The spikes are passed through a low-pass filter, which integrates them into a video waveform. This video signal still contains the pre-emphasis applied in the record process and must be de-emphasized before application to the output.

De-Emphasis

The high-frequency components of the video signal must be de-emphasized before the signal is fed to the output of the machine.

133

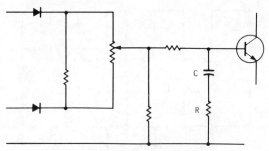

Fig. 7-26. De-emphasis network.

This is accomplished by a simple RC series network (R and C in Fig. 7-26). This can be found either in the video output amplifier or before the low-pass filter that integrates the spikes from the demodulator.

PIM Demodulator

The playback of a pim waveform involves the same sequence of events as in other systems; the signal is taken from the head, amplified in a head preamplifier, passed through a limiter chain, and then demodulated. The limiters have a push-pull output to ensure symmetry of the pim output waveform. This output is differentiated so

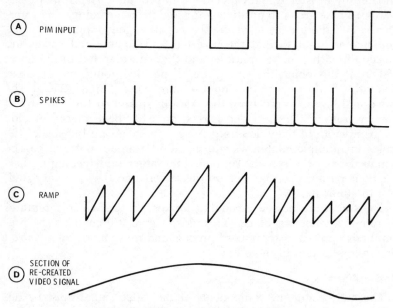

Fig. 7-27. Waveforms in pim demodulation.

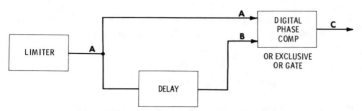

Fig. 7-28. Digital method of demodulation.

that at each zero crossing of the pim waveform, a narrow pulse is produced. These pulses are passed through a full-wave rectifier to give a train of positive-going pulses (waveform B, Fig. 7-27), which is applied to the base of a transistor. The collector circuit of the transistor contains an RC network that generates a series of constant-slope ramps (waveform C, Fig. 7-27). A low-pass filter averages the area of these ramps to produce the output video signal (waveform D, Fig. 7-27). The video signal is amplified and de-emphasized to give the standard NTSC signal, which is routed to various places within the machine and to the output connectors for display on a monitor.

The fm signal can also be demodulated by digital methods. After limiting, the fm square wave is split into two paths (Fig. 7-28). One is a direct path, and the other passes through a delay circuit. The output of the delay forms one input to the next circuit, and the direct path forms the second input. The delay can be produced either by a passive phase-shift network or by an injection-locked monostable multivibrator. The final circuit is a digital phase comparator or an Exclusive-or circuit.

In Fig. 7-29, waveform A is the fm signal, and waveform B is the delayed signal. The Exclusive or gate will produce an output pulse only when one input or the other is positive. Its output (waveform C) will stay low when both inputs are of the same polarity. The

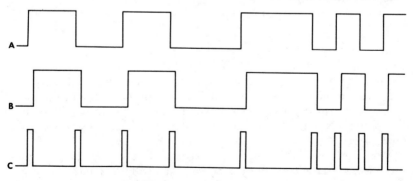

Fig. 7-29. Waveforms in digital demodulation.

resulting train of pulses is then converted into a video signal. As before, the pulse train is at twice the fm carrier frequency and thus is outside the video bandwidth.

ELECTRONICS TO ELECTRONICS

Electronics to electronics (E-E) is a facility peculiar to vtr's. In a professional audio recorder, a separate playback head and amplifier make it possible to check the recording a fraction of a second after it has been put on the tape, thus ensuring quality and the presence of a signal. The smaller, less expensive recorders require the tape to be checked after the recording is finished.

Video recorders are in a sort of halfway position. During recording, the signal viewed on the monitor plugged into the vtr has passed through all of the record and playback electronics. It is picked off

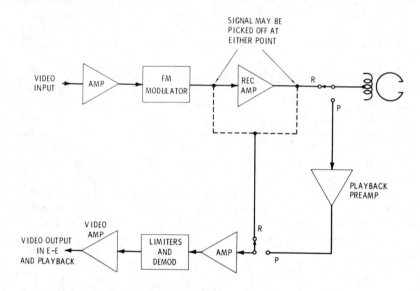

Fig. 7-30. Electronics-to-electronics operation in a vtr.

the record chain just before it is applied to the heads, and it is then fed through the playback process. Thus the performance of the electronics in the machine can be checked readily. To determine whether the picture is actually on the tape, the recording must be played back later. Fig. 7-30 demonstrates the E-E process.

Most portable machines do not have an E-E capability. The picture viewed in the camera eyepiece is taken from an emitter follower just after the first-stage agc in the record amplifier.

In some recent larger machines, a flying playback head has been mounted in the scanning disc. It is about 120° behind the main recording head and mounted so that it scans the track that has just been recorded. The output of this head is fed to its own playback preamplifier and then through the normal playback or E-E chain to give a video output. In this way, the actual recording on the tape can be monitored. The manufacturers stress this is not a quality picture, but merely an indicator that the recording has been made on the tape.

8

Servos

In audio, the precise occurrence of one event in the program compared to any other event is unimportant in the engineering sense. In video this is not so. One of the main characteristics of the video signal is the regular occurrence of the sync pulses, and these must be recorded so that they can be reproduced in the same form and timing relationships that they had in the original signal.

In nonbroadcast vtr's, it is only necessary to pay attention to the timing of the vertical interval to record and play back a satisfactorily stable picture. In order to record the video signal in a definite pattern on the tape, the heads must be made to assume a definite physical position with respect to the tape at the time of the arrival of the vertical interval. This positioning and timing is brought about by the use of a servo system.

Servo Fundamentals

A servo is simply an electronic or mechanical arrangement that permits self or automatic control over the speed and positioning of some piece of machinery. To operate properly, a servo must have two pieces of information. First, it must know what the machinery is doing, so some sensing device on the output is needed. Second, it must have some constant information that tells it how far the machinery has strayed from its desired state. Thus the two inputs to a servo are a feedback signal and a reference signal. These are fed to a comparator circuit, which produces an error signal proportional to the difference between the two inputs. This error signal is then used to correct the speed and position of the machinery. Control is fully achieved when the machine stays within predetermined limits.

Servos can be made to be extremely accurate and exert a very fine control over a mechanism. Two common examples are the automatic pilot on an aircraft and the hydrofins on an ocean liner. One of the most exacting uses of servos is in the modern vtr. To give some idea of the accuracy achieved, a displacement of one tenth of a line in a tv picture is objectionable. This is about 4 minutes of arc in a rotating head, or about 6 microseconds of error.

In vtr's, the rotating heads are controlled by a servomechanism on both record and playback so that the head is in its correct position with respect to the vertical interval in the record mode, and with respect to the signal on the tape during playback. A capstan servo is not required in the simplest machines; it is sufficient to drive the tape at a constant speed with a minimum of flutter and wow. A capstan servo becomes necessary when editing is required, and it is also needed in portable machines, but for a different reason.

The servo circuit in vtr's is required to do other things in addition to positioning the heads. The manufacture of rotating machinery is not perfect, and the tolerances in normal production can cause slightly eccentric capstans, minor misalignments of the head-drum assembly, and slight imperfections in the placing of the various parts that make up the whole vtr. When tension changes and stretch in the tape are added, the result can be an unusable picture. The servo corrects for these imperfections and aids in producing near perfect recording and playback of the tv signal.

In recording a tv signal, five basic possible sources of signals exist: (1) off-the-air tv signal, (2) a camera with internal drives or driven from a sync generator, (3) another vtr, (4) an input from any of the preceding three sources for editing purposes rather than straight recording, and (5) a camera driven from the vtr. In the first four of these modes, the one thing in common is that a video signal is fed into the vtr to be recorded. In the first three cases, the signal is treated exactly the same. In mode 4, a capstan servo is required because there are two video signals that must be matched in time. In mode 5, the vertical (and sometimes the horizontal) signal is developed in the vtr from the rotating head mechanism, and this causes a slightly different mode of servo operation.

TYPES OF SERVO IN COMMON USE

Although there are many different types of servomechanisms, those used in vtr's are limited to a few common types. A basic block diagram of a typical vtr servo system is shown in Fig. 8-1. The blocks are marked according to their function and their type of output sig-

Chart 8-1. Parts of Servo System

Inputs	Comparator Circuit	Comparator Output	Control Circuit	Control Output	Driver	Mechanism
Vert sync	Ramp Sampler	DC	DC Amp	DC	DC Amp	Braking Coil
Tach Pulse	MV Duty Cycle	Square Wave With Variable Duty Cycle	VCO	Varying AC Freq	VCO	DC Motor / Synchronous AC Motor
Control Pulse	Sliced Ramp	Pulse Width	Blocking Osc	Pulses	DC Switch	Nonsynchronous AC Motor
Frequency Generator	MV With Added Pulse					DC Subsidiary Motor

nal. Chart 8-1 lists the types of circuit or mechanism most usually found in these blocks.

All vtr's follow the basic layout shown in Fig. 8-1, and this arrangement is common to both head and capstan servos. The details within the blocks can be any of the forms shown in the chart. When a certain choice is made for one block, it does not follow that this will determine the other blocks. For example, a vertical-sync input does not have to be fed into a ramp sampler circuit and then this used to control a dc amplifier. The vertical-sync pulse could feed a multivibrator, and the varying-duty-cycle output of the multivibrator could be used to trigger a blocking oscillator. Several different choices are found in practice, even within the various models from the same manufacturer. Some choices are peculiar to one maker only,

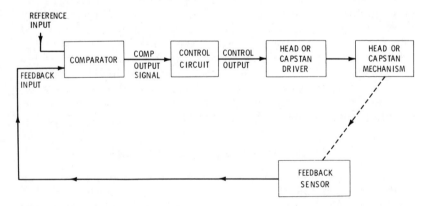

Fig. 8-1. Basic servo block diagram.

and others are universal. Also, the circuit configurations of similar blocks can differ widely from machine to machine.

The following sections show some of the more common methods used and the basic circuit configurations. In some cases, the functions of comparison, control, and drive are not clearly separated, and often one circuit will be used for more than one purpose or will be missing entirely. After analysis of the separate sections, some typical vtr servo circuits will be shown. All the examples have been taken from vtr's in current commercial use.

INPUT SIGNALS

There are five input signals in common use in a vtr:

1. The vertical-sync pulse, stripped from the incoming video
2. A head-tach pulse, derived from the rotating heads
3. A control-track pulse, which has been recorded along one edge of the tape
4. A 60-Hz sync-pulse signal derived from the ac line instead of the incoming video signal
5. A tone generated by either the rotating heads or the capstan motor

The vertical-sync pulse is always used as a reference pulse in the servo circuits, for both head and capstan, whereas the others are generally used as feedback information, since they have been derived from the moving parts of the system.

Considerable care is taken in the design of the circuits that process these signals. To eliminate false servo operation caused by noise and spurious signals, almost all of the inputs are converted to square waves of various duty cycles. This is done by using the input as the trigger pulse for a one-shot multivibrator.

Vertical-Sync Pulse

After the input video signal has passed through an agc section, it is fed to a sync-separator circuit on the servo board. This circuit performs the same function as the sync separator in a conventional tv set. Its output is a vertical-sync pulse, which can be either positive going or negative going, depending on the design of the following circuit. Fig. 8-2 is a simplified version of the sync stripper and vertical-pulse separator used in the Sony AV 3650.

Head-Tach Pulse

The purpose of the head-tach pulse is to indicate the speed and position of the rotating heads. It is produced all the time the heads

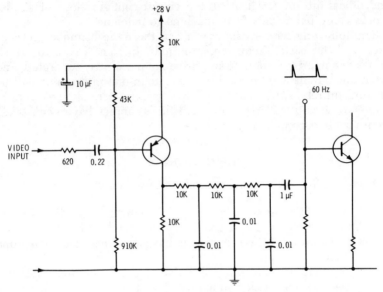

Fig. 8-2. Sync stripper circuit.

are rotating, and it is used for several different purposes in the vtr. One-head machines produce one pulse per revolution of the head drum, and two-head machines produce two pulses. One pulse is sufficient for head rotation control, but in the two-head machine the tach pulses are also used for head switching.

Two methods are generally used to produce the head tach pulses:

1. A magnetic pulse. This is usually produced by a metal vane or a small magnet mounted on the rotating mechanism; the vane or magnet passes over or under a coil or small pickup head similar to the heads used in an audio machine.

2. A light is made to reflect into a phototransistor for a brief instant by a reflective strip mounted on the rotating mechanism.

Each method produces a strong pulse and gives reliable operation, but the magnetic method is the most common.

Sony AV 3400—This method is used universally in Sony vtr's: Two coils are mounted above the rotating plate that carries the video-head bar. This plate has a vertical metal vane mounted at right angles to each head and passes very close to the two coils. The coils are at opposite ends of a diameter of the head drum and are connected as shown in Fig. 8-3. They are biased about 2 volts positive, and, as the vane passes, a negative pulse is produced. This triggers a bistable multivibrator, the output of which is a 30-Hz square wave.

142

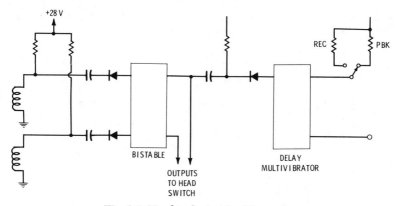

Fig. 8-3. Head-tach circuit of Sony vtr.

Both outputs of the bistable are fed to the head-switching circuit in the playback preamplifiers, but only one output is used for the servo. This output is differentiated by a capacitor and resistor, and the negative spike that results is used to trigger a monostable multivibrator. The width of the output of this device is preset by potentiometers to give the correct output for both recording and playback control, and the negative-going edge is used to trigger the comparator gate circuit.

Ampex—With only one head, the Ampex machines need only one tach coil to produce a single pulse for each revolution. The pulse is produced by a magnet embedded in a rotating brass plate below the head scanner assembly. A small coil is mounted close to the plate, and a pulse is produced every time the magnet passes the coil. The output of the coil is ac coupled to a normally off transistor to produce a negative-going pulse of about 0.2 to 0.3 volt, which is further processed by the servo circuitry.

The circuit is arranged so that an absence of a tach pulse causes the head motor to run fast. In this way, it very quickly comes up to speed for servo lock-up, and this also prevents motor hang-up at low speeds. Fig. 8-4 shows how the pulse is amplified.

IVC—In contrast to most other machines, the IVC series of vtr's use a phototransistor to generate the tach pulse. A small stationary lamp is mounted above the head scanner plate, and next to it is mounted the phototransistor. Shielding is provided so that light from the lamp does not reach the transistor directly. On the rotating plate is a strip of metal that reflects the light from the lamp into the transistor when the strip passes beneath them. Thus an output pulse is produced once per revolution of the head.

Fig. 8-5 is a simplified block diagram of the circuit. The tach pulse gates a portion of the ramp to give a dc level into the capacitor. If

143

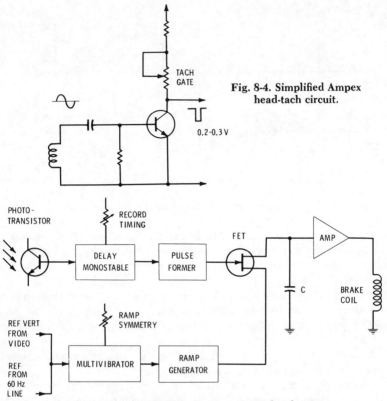

Fig. 8-4. Simplified Ampex head-tach circuit.

Fig. 8-5. Simplified block diagram of IVC head servo.

the pulse is early or late, a different dc level is fed to the capacitor, and this controls a braking action to correct the head rotation.

Control-Track Pulse

When a tape is played back, the rotating heads must be positioned so that they track the recorded signals exactly. To be able to do this,

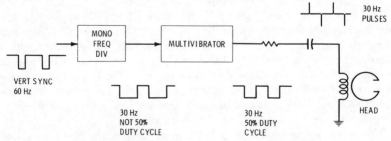

Fig. 8-6. Block diagram of control-pulse recording.

it is necessary to know where the recorded tracks begin their sweep across the tape. For this purpose, a series of pulses are recorded along one edge of the tape. These are derived from the incoming vertical sync, and one pulse per revolution is laid down; thus a one-head machine has control-track pulses at a 60-Hz rate, and a two-head machine has them at a 30-Hz rate. These pulses have been compared to the sprocket holes of a film, and they serve about the same purpose. Due to the placing of the control-track recording head, the pulses bear a definite relationship to the beginning of the slanted video tracks, and their respective placement can be seen quite clearly if the tape is developed.

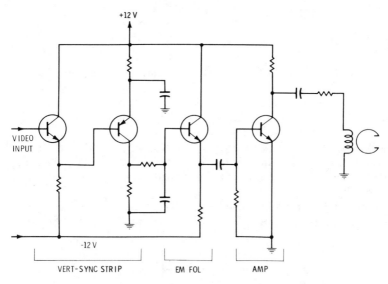

Fig. 8-7. Simplified Ampex control-pulse record circuit.

Recording the Control Track—A typical arrangement is shown in Fig. 8-6. The 60-Hz vertical-sync frequency is divided by a mono-stable with an on time long enough to ignore every other pulse. This output is fed to a multivibrator that converts the signal to a 30-Hz square wave with a 50-percent duty cycle. The square wave is then differentiated by an RC network and applied to the head so that pulses are recorded along one edge of the tape.

In the circuit shown in Fig. 8-6, which is typical of two-head machines, the first monostable has a pulse duration of about 24 milliseconds. This allows it to ignore alternate incoming pulses and produce a 30-Hz output. This output does not have a 50-percent duty cycle, and it is used to trigger another monostable that does have a

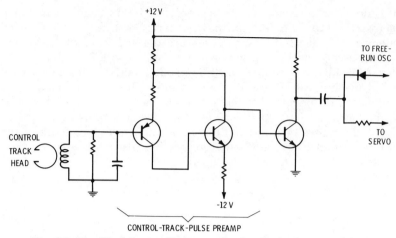

Fig. 8-8. Simplified Ampex 5000 control-pulse playback preamplifier.

50-percent duty cycle. Thus the 60-Hz incoming vertical-sync pulses are now reduced to a 30-Hz square wave. This is differentiated and applied to the control-track record head.

One-head machines record the control track at 60 Hz, as in the circuit in Fig. 8-7, which is taken from the Ampex 5000. The amplifier conducts only on the positive tips of sync, thus providing noise immunity.

Control-Track Playback—On playback, the control-track head is connected to a simple pulse-amplifier circuit that amplifies the in-

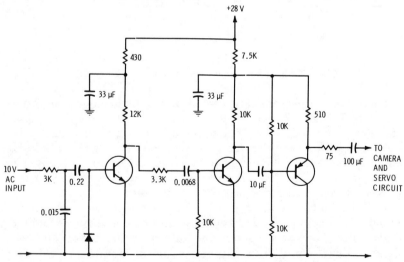

Fig. 8-9. Sony AV 3650 vertical drive circuit.

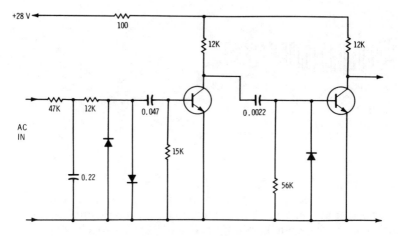

Fig. 8-10. Sony EV 320F vertical drive circuit.

coming pulse to produce a clean, noise-free pulse suitable for triggering the following multivibrators. The pulse outputs of the multivibrators are used by the head servo on playback as a reference signal to position the heads. The control signal is of vital importance to the correct playback of a recorded tape; if it is missing, correct tracking cannot occur, and picture stability is lost. A basic version of the Ampex 5000 control-track playback preamplifier is shown in Fig. 8-8.

60-Hz Pulse Formed From Power Line

In the absence of an incoming video signal, pulses occurring at 60 Hz can be formed from the power line to replace the incoming vertical sync. This procedure is only necessary in a machine equipped for editing, where an input reference signal is continually required for servo lock-up. Two typical circuits are shown in Figs. 8-9 and 8-10. These simply clip and amplify the incoming 60-Hz sine wave until a square wave is produced. The square wave is differentiated to give a series of pulses at a 60-Hz rate.

Motor-Generated Tone

In some machines, a toothed wheel (Fig. 8-11) is incorporated in the capstan motor or the rotating head mechanism. A definite number of teeth on the wheel pass close to a coil mounted on a pole piece. The result is an output with a frequency determined by the number of teeth and the number of revolutions per second of the motor. This output signal is used for two main purposes. It can be a servo feedback signal, or the frequency can be made 15,750 Hz so that it can serve as the horizontal drive for a camera.

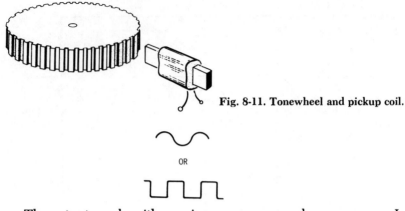

Fig. 8-11. Tonewheel and pickup coil.

OR

The output can be either a sine wave or a rough square wave. In either case, the signal is amplified and shaped to eliminate noise before being fed to subsequent circuits.

COMPARATORS

The function of the comparator is to accept the processed input signals and to compare them. From this comparison, an output signal proportional to the difference between the inputs is produced, and it is used to control the head or capstan position and rotation speed. Whichever inputs are compared, each will have first been fed to some form of multivibrator to produce a conditioned noise-free signal that is of suitable amplitude and in an approximate time relationship to the other signal.

Two basic forms of comparator circuit are widely used:

1. A ramp waveform sampled by a pulse on either a positive or negative slope
2. A ramp sliced at some point to produce a pulse of varying width

The first method is the most popular. Other methods, such as varying the duty cycle or symmetry of a square wave, have been used, but never to the extent of the first method.

It is sometimes difficult to decide just where the comparator begins and what is correctly a signal processor prior to comparison. Here the signal processing is treated as part of the overall comparator circuit for simplicity.

Ramp Sampling

Ramp sampling is probably the most common form of comparator found in the modern helical vtr. The basic circuit block diagram is shown in Fig. 8-12. The reference input is integrated to a ramp, which is applied to one input of a gate. A pulse formed from the feedback input is applied to the other gate input. The result is that the ramp is sampled at some point on its slope, and this causes a gate output with a duration determined by the pulse duration and an amplitude determined by the sampling point on the slope. The output pulses are then integrated to a dc level and applied to a dc amplifier or are used to trigger an oscillator of some form. (In some machines, the feedback pulse is integrated to form the ramp, and the reference pulse is used as the sampler. Either method works well.) Variations in the arrival time of either pulse will cause the ramp to

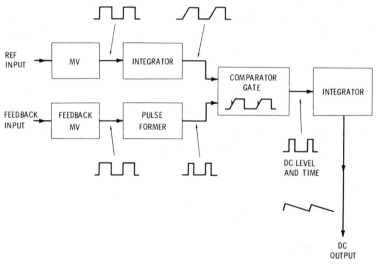

Fig. 8-12. Block diagram of ramp sampling.

be sampled at a different point and thus change the dc output level. This dc level is used to correct the rotating mechanism until the sampling occurs at the correct place for servo lock-up.

Most machines use the same ramp for both record and playback, but for playback it is made less steep. Some models use the positive slope for record and the negative slope for playback. On record, a steep slope is needed so that the slightest deviation will produce a quick servo response to get a quick reaction and correction. This helps to minimize control-track variations recorded on the tape. On playback, a longer slope is used to provide a slightly slower servo response time. This absorbs control-track variations caused by wow and flutter and gives a smoother playback with less continual hunting by the servo.

Fig. 8-13 shows how changing the size of a capacitor in the Sony AV 3600 changes the slope of the ramp when going from record to playback. In each case, the sampling is on the negative-going slope and occurs at the same dc height above ground to give the same dc drive to the motor brake; hence, the sample pulse is delayed for a different time in each mode.

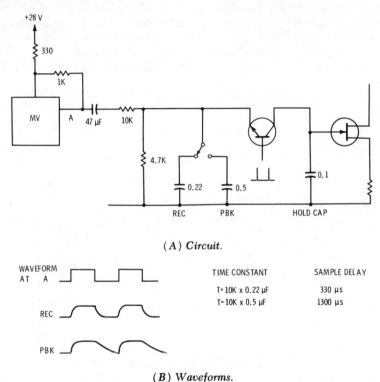

(A) Circuit.

WAVEFORM AT A	TIME CONSTANT	SAMPLE DELAY
	T = 10K x 0.22 µF	330 µs
	T = 10K x 0.5 µF	1300 µs
REC		
PBK		

(B) Waveforms.

Fig. 8-13. Ramp-slope circuit of Sony AV 3600.

Fig. 8-14 shows how the Sony AV 3600 uses the positive-going slope on record and the negative-going slope on playback. The arrival time of the sampling pulses remains constant at 900 microseconds after the flip-flop changes state.

An unusual sampling circuit is used in the Shibaden SV-700 (Fig. 8-15). The ramp is passed through a diode bridge and is sampled by the vertical-sync or control-track pulses. The output is a pulse the height of which is determined by the position on the ramp at which the sampling occurred. The output is integrated to a continuous dc by a capacitor. The same slope and sample time are used in both record and playback.

150

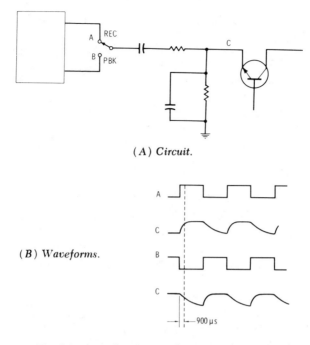

(A) Circuit.

(B) Waveforms.

900 μs

Fig. 8-14. Use of positive and negative ramp slopes.

Ramp Slice

In the ramp-slice method, the input signal is used to produce a square wave which is then integrated into a ramp waveform. The ramp is compared to a dc level, which can be either preset or produced by another input. When the ramp reaches the dc level, a gate is made to produce an output pulse, the duration of which is determined by the time it took the ramp to reach the dc level. Hence, pulse duration can be controlled by either the ramp or the dc level.

The following circuits show how two slightly different forms of ramp are generated by an input pulse. In Fig. 8-16, Q2 is normally held off by the 1K resistor, allowing capacitor C to charge linearly with a time constant T1 determined by the 130K resistor and the 30K potentiometer.

When the ramp reaches 2 volts, the differential amplifier changes state to produce the negative output pulse. The square-wave input to Q1 is from a coil in the capstan motor. This is differentiated into pulses that are applied to the base of Q2. The positive pulse turns Q2, on, allowing the capacitor to discharge very fast with a time constant T2 determined by C and the saturation resistance of Q2. Variations in the motor speed alter the time at which transistor Q2

151

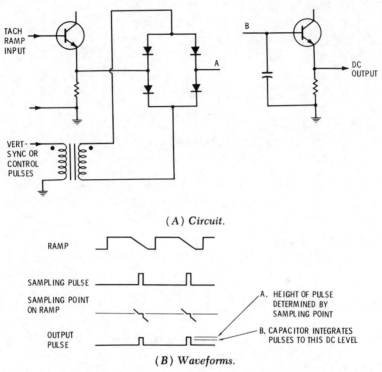

(A) Circuit.

RAMP

SAMPLING PULSE

SAMPLING POINT
ON RAMP

OUTPUT
PULSE

A. HEIGHT OF PULSE
DETERMINED BY
SAMPLING POINT

B. CAPACITOR INTEGRATES
PULSES TO THIS DC LEVEL

(B) Waveforms.

Fig. 8-15. Pulse sampling in Shibaden SV 700.

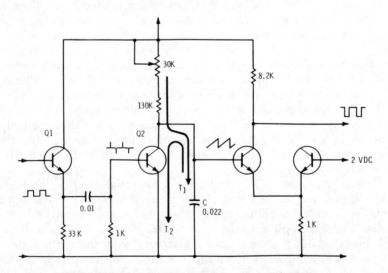

Fig. 8-16. Ramp generator circuit with fast discharge.

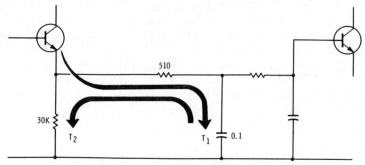

Fig. 8-17. Ramp generator circuit with slower discharge.

conducts and thus alter the width of the output pulse from the differential amplifier. The output pulses are used to drive the motor-drive circuit.

A slightly modified version of this circuit is shown in Fig. 8-17. Here the capacitor is charged from the emitter circuit of the transistor, so on discharge a longer time is required, thus producing a longer and less steep ramp. Fig. 8-18 shows the time relationships of the waveforms in such a circuit. Varying the spacing of the input

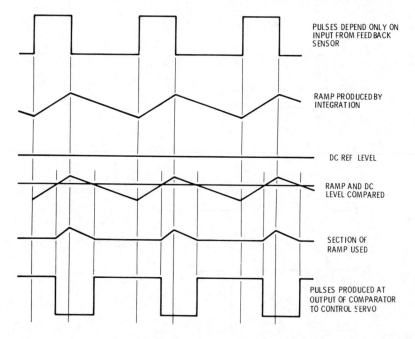

PULSES DEPEND ONLY ON
INPUT FROM FEEDBACK
SENSOR

RAMP PRODUCED BY
INTEGRATION

DC REF LEVEL

RAMP AND DC
LEVEL COMPARED

SECTION OF
RAMP USED

PULSES PRODUCED AT
OUTPUT OF COMPARATOR
TO CONTROL SERVO

Fig. 8-18. Ramp and pulse timing relations.

pulses or the dc level will vary the section of the ramp used to produce the output pulse, so either the reference or the feedback input can be used to control the servo. This circuit is a simplified form of the capstan speed servo employed in the Sony AV 3650.

FG-Type Comparator

An interesting type of speed-control comparator is found with motors that contain a frequency-generator (FG) coil. The output

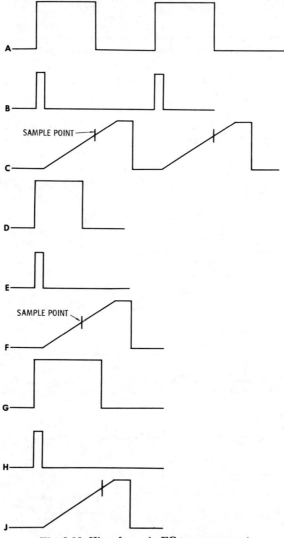

Fig. 8-19. Waveforms in FG-type comparator.

from the coil is shaped into a square wave, and the leading edge is used to trigger a constant-width pulse generator. In Fig. 8-19, waveform A is the square wave, and waveform B is the constant-width pulse. The trailing edge of pulse B is used to start a ramp, and the ramp is sampled by the falling edge of waveform A. The sample point is indicated on the ramp in waveform C.

If the motor runs fast, the frequency of the square wave increases, and the input pulses are shorter (waveform D). These pulses still produce the constant-width pulse (repeated as waveform E for convenience), which produces the same ramp (shown again as waveform F). The falling edge of waveform D now samples the ramp earlier, as indicated on waveform F, and a lower dc level is produced at the comparator output to reduce the motor speed.

If the motor runs slow, the longer input pulses appear as in waveform G. The pulse and ramp are produced exactly as before (waveforms H and J), but now the falling edge of the input pulse samples the ramp at a higher level. This gives a higher dc output to the motor.

When this method is used, very small changes in motor speed will produce an immediate reaction. Obviously, the input square wave is the feedback information from the motor, but what is not so obvious is that the reference in this type of comparator is the constant-width pulse, shown here as waveforms B, E, and H. What is unusual about this circuit is that it has only one input, which accepts the feedback signal. The reference input is missing because the reference signal is contained inside the circuit. This circuit is used in several modern vtr's and is contained entirely within one IC.

Comparator Output and Integrator

Usually, the output signal of the comparator is applied directly to the control or driver circuit without any further processing. The main exception to this occurs when the comparator output is a pulse and the next stage requires a dc input. In this case, the pulse is integrated to a dc voltage, the level of which depends on the pulse width and pulse spacing.

A typical circuit is shown in Fig. 8-20. In this case, a ramp waveform is applied to the emitter of Q1, and a sampling pulse is applied to the base. The collector waveform (at point D) is a series of pulses. These pulses charge the capacitor, which then holds its charge due to the long time constant presented by the high input impedance of the FET used for Q2. This causes the integration of the pulses to a dc level, which sets the dc level at the output cf Q2. It is this dc level that is now used for further servo control. Variation in the arrival time of either the pulse or the ramp causes sampling at a different time and hence a varying height of the pulse at point D, thus causing the output dc level to vary. (The waveform shown at point D is

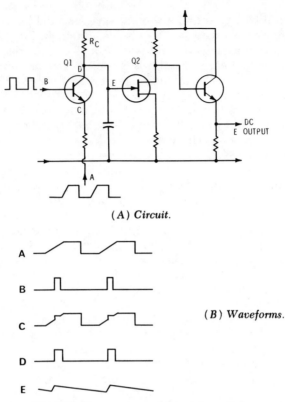

(A) Circuit.

A

B

C   (B) Waveforms.

D

E

Fig. 8-20. Comparator and integrator action.

the waveform that would exist if point D were disconnected from point E. The waveform that results from integration is shown at point E.)

CONTROL CIRCUITS

The control-circuit stage accepts the output of the comparator and uses this signal to develop the control signal for the driver of the rotating mechanism. Various circuits have been used, but a dc amplifier is the most common.

Although many different configurations of dc amplifiers are found, those used in vtr servos have one purpose only. This is to amplify an input current or voltage to a level sufficient to drive a mechanical device, usually a motor or a braking coil.

The dc amplifier is in a gray area of control section and output driver combined, since it is difficult to separate either the functions or the stages. The examples shown in Figs. 8-21 and 8-22 are basic

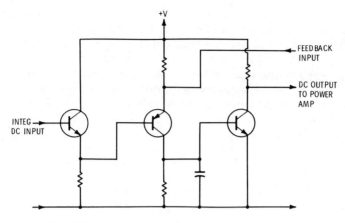

Fig. 8-21. Circuit of dc amplifier from Ampex 5000.

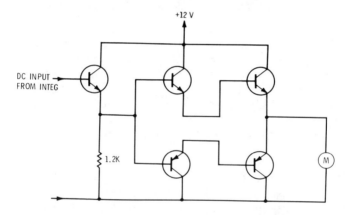

Fig. 8-22. Circuit of dc amplifier from Sony EV 320.

circuits of dc amplifiers. Many recent machines use a simple IC op-amp.

Fig. 8-22 is a case of combined functions. The motor is a subsidiary motor and not the main head or capstan motor. It uses a belt drive to provide capstan servo action.

ROTATING-MECHANISM DRIVERS

The purpose of the driver is to apply the control voltage or current to the head or capstan mechanism and to provide the control for that mechanism. A dc power amplifier is used almost universally for this purpose. As was pointed out in the last section, it is not always possible to distinguish this stage from the previous control section.

The input to a dc amplifier in a vtr servo is a slowly changing or nearly constant dc level. Often, it is the output from the IC op-amp that follows the comparator. The current input is changed slightly by the servo control circuit to provide correction, and a small voltage change at the input and output can be observed with an oscilloscope when the circuit is functioning properly. The amplifier is usually composed of one or two stages of dc amplification, mainly to provide current amplification; hence, emitter-follower and Darlington configurations are quite common. Feedback is sometimes incorporated over the whole stage, and it can even be applied to the input of the control stage from the final output to the motor. The purpose of this is to control hunting, and often a different time constant is used for record and playback. A simplified diagram of the Sony EV 320 motor-drive amplifier is shown in Fig. 8-22.

ROTATING MECHANISMS AND CONTROL DEVICES

The two rotating mechanisms that must be controlled are the heads and the capstan. Although a multiplicity of methods have been used, most small helical vtr's now use a dc motor. Often, this is the brushless type with a printed armature. In many cases, the motor provides a belt drive to the rotating video-head drum, and a magnetic brake is used to control the speed. In some machines, a subsidiary dc motor is used with a belt drive. Where servo control is required, ac motors are no longer in common use.

DC Motors

Both capstan and head control systems may use dc motors, but this type of motor tends to be more common in capstan servos. The most usual drive method is by belt, the other methods being used rarely. The electronic drive signal is most often a dc level, but square waves, pulses, and switched dc voltages have been used with varying degrees of success. Recently, vtr's have begun to incorporate brushless and semiconductor motors, for which greater reliability and ease of accurate control are claimed.

Several types of dc motors have been constructed with a toothed wheel and pickup coil as an integral part. This has been used both for servo feedback control and for camera drive signals.

Eddy-Current Braking

The principle on which eddy-current braking works is quite simple. If a motor is rotating too fast, the easiest way to slow it down is to apply a simple brake pad or one's finger to the shaft or wheel so that friction will slow the rotational speed. Varying the pressure of the brake pad will effect some control over the speed. Because mechanical brakes are somewhat imprecise and tend to wear out, their use would be somewhat limited. But a more sophisticated version of this idea is available. If a coil of wire with a current through it is placed near the rotating part, it will act like an electric motor and tend to produce a rotation of its own. If this rotation is arranged to be in the opposite direction to the main rotation, a slowing action will occur. Varying the current in the coil will vary the magnetic force around the coil and thus will vary the braking effectiveness. All of this occurs without any mechanical contact. The braking coil acts like a squirrel-cage motor in reverse; i.e., the rotor turns in a stationary field, and interaction between the field induced in the rotor and the static field produces the braking action, which is directly proportional to the current in the coil. This principle has been used in the speed control of turntables in broadcasting for some time and has proven to be a successful method of speed control of the heads in vtr's as well.

In a vtr, the head drum is usually belt driven from a synchronous motor with the pulley ratio such that the head runs about 1 percent fast. Its position and speed are then sensed by the head-tach coils or reflective strip, and the servo output is applied to the braking coil to slow the motor until phase and speed correction are achieved.

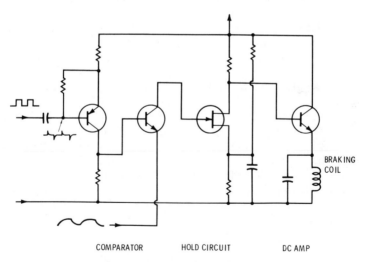

Fig. 8-23. Driver for eddy-current brake.

A typical circuit for a braking coil is shown in Fig. 8-23. It is merely a simple dc amplifier driven from a comparator and integrator. About 25 milliamperes of current is a common value found in most braking coils of this type.

THE USE OF MULTIVIBRATORS IN SERVO CIRCUITS

When the main input signals—the head-tach, vertical-interval, and control-track pulses—are fed into the electronics, they are first amplified and then used to trigger monostable or other types of multivibrators. The outputs of these multivibrators are thus delayed in time from the actual occurrence of the input pulses. There are three main reasons why this processing is required:

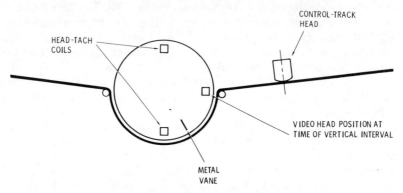

Fig. 8-24. Head-tach vane and coils.

1. To eliminate noise
2. To correct for mechanical imperfections
3. To aid in their mutual alignment for comparison prior to control

The pulses from the head occur when a vane of metal passes near a pickup coil (Fig. 8-24). Due to the mechanical structure of the head drum, the coils can be placed in several positions, but in general the placement is such that the pulses are nowhere near coincidence with either the incoming vertical sync on record or the control-track pulses on playback. This difference in timing is not absolute and can vary considerably with mechanical imperfections. The replacement of these parts during repairs can also further disturb the mechanical alignment. It does not matter how far apart in time these pulses are as long as the difference remains constant. Then minor variations in this time difference can be used as a control for the servo.

If the head-tach pulse is used as a reference marker, the vertical-sync interval and the control-track pulse occur at different times, and since each is compared with the head tach for record or playback control, they must obviously be delayed by different times. To achieve a time comparison of these pulses, they are artificially moved closer by the use of multivibrators. This enables a fine control to be incorporated to compensate for mechanical imperfections and to get rid of noise, as well as achieve electronic time comparison. The minor time variations can be made to assume major importance, and thus very tight control of the servo is achieved.

SPECIAL INTEGRATED CIRCUITS

Recently, special ICs have begun to be incorporated into vtr's. Where a relatively simple circuit consisting only of transistors and resistors is required, it is now a simple matter to manufacture an integrated circuit. Custom ICs are now well spread throughout the electronics industry, and many manufacturers are using them. They have the advantages of space-saving size, ease of servicing, and low cost in large quantities. In many cases, it is cheaper to have an IC made than to assemble parts to a printed circuit board. A simple example from the servos in the Sony AV3650 is shown in Fig. 8-25.

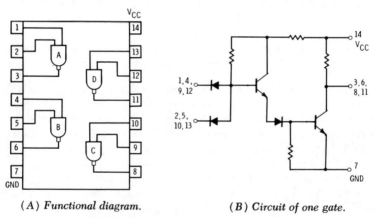

(A) Functional diagram.   (B) Circuit of one gate.

Fig. 8-25. Example of an integrated circuit.

The Head Servo

Every vtr needs a head servo. There is no way in which a tv signal can be recorded or played back without controlling the rotation of the heads, and so a servo must be used in both modes. The reasons for using a head or drum servo can be summarized as follows.
   In record:

1. The video must be positioned at a precise spot with respect to the tape at the time of arrival of the vertical interval. This permits a definite recording pattern to be established on the tape.

161

2. The timing of the head switching or the dropout position must be accurately adjusted to its chosen location, and this must be maintained throughout the recording.
3. The dc level, duty cycle, or frequency of the motor drive signal must be maintained within close tolerances at all times to provide an accurate drive.

In playback:

1. The video head must be positioned at the same spot on the tape as the recorded video tracks, and it must maintain the beginning of this tracking in synchronism with the control track.
2. Head position and phase sensing must be possible.
3. The tracking errors must be adjustable to provide correct tracking of the recorded signal.
4. The effects of wow, flutter, tape tension, and tape stretch must be minimized in the mechanism that drives the tape.
5. The drive signal to the head motor must be controlled to provide the above conditions.
6. Correct head switching and dropout must be effected, and they must coincide with those on recording.

Regardless of whether the head scans the tape from top to bottom or vice versa, it must begin and end its scan with a high order of accuracy and repeatability. As the head rotates, its position and speed are sensed by the head-tach devices, and in this manner a definite pulse is produced each time a head passes some definite point. One-head machines have one sensor to produce one pulse per revolution, whereas two-head machines have two pulses per revolution. One pulse is sufficient for servo control, but two are required for correct head switching and for vertical-interval insertion on slow-speed and still-frame playback.

Basically, the object of servo control is to maintain the correct spatial relationship between the head and the tape. The head drum is usually belt driven from an electric motor, but direct drive is sometimes used.

The same circuit, with minor changes, is used in both the record and playback modes. In the ensuing paragraphs, record is examined first, and then playback. Examples from actual machines are used.

RECORD MODE

All servos for head control use the same two signals in the recording process. The incoming vertical-sync interval is used as the reference, and the pulses from the rotating heads are the feedback information. These two signals are first processed by the usual multivibrators to provide a clean, noise-free pulse of some definite width and amplitude. These signals bear a definite time relationship to the incoming signals, and this is maintained continually. They are then compared by a comparator circuit to produce an error signal for controlling the head rotation. The object of the control signal is to place the heads so that they assume their correct position with respect to the vertical interval, and thus begin the tape scan at the correct point in time and at the correct position on the tape. Any variation of either time or position will displace the head pulse with respect to the vertical-interval pulse and thus cause the comparator to produce a different output signal. This output signal acts to control the head rotation in a direction to correct the timing relationship. Fig. 8-26 is a diagram of a typical head servo in the record mode.

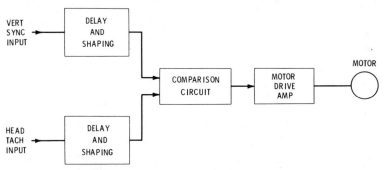

Fig. 8-26. Basic block diagram of head servo in record mode.

PLAYBACK MODE

Head servo action in the playback mode is essentially the same as in the record mode, but with two major differences. First, the reference input signal now comes from the control track on the tape, and second, a tracking control is inserted into the control-pulse chain prior to the comparator section. The head-tach pulse remains as the servo feedback information. Fig. 8-27 shows a simple playback-servo block diagram that serves as a basic model for all helical-vtr servos.

Tracking refers to the ability of the head to scan correctly the recorded tracks in the playback mode, and thus reproduce a perfect picture. Ideally, a tracking control should not be necessary, and on a perfectly aligned machine it is not used. But a stretched tape or a control head slightly out of alignment will cause a minor mistiming in the arrival of the control pulses to the servo, and this will make the heads track slightly away from the center of the recorded tracks. The remedy is to delay the control pulse electronically, and this has the same effect as moving the head sideways to secure the correct timing. Because minor variations occur in manufacture due to mechanical tolerances, this minor misadjustment of the head does happen and thus must be corrected.

The usual method is to insert an extra multivibrator in the circuit; the pulse width of this multivibrator can be varied by a control placed on the front panel. The operation of the control varies the position of a ramp or pulse in the comparator circuit, thus varying the control on the head motor.

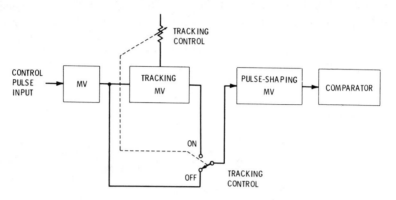

Fig. 8-27. Block diagram of head servo in playback mode.

On a two-head machine, about 12 milliseconds of variation are needed, and on a one-head machine about 25 milliseconds are required. This is why the one-head machines have two tracking multivibrators and two ganged potentiometers as a tracking control.

The physical distance of the control head along the tape from the point where the video head begins its scan is of critical importance to correct playback. If this distance is incorrect, the head will not follow the recorded tracks across the tape. Minor differences in head placement can be taken care of by the tracking control, but a major error—such as when a new head is installed—is quite likely to be outside the range of the tracking control. In this case, the correct placing of the head can easily be achieved, since there is provision for lateral adjustment in the head mounting. The easiest way to effect positioning is to play back a standard alignment tape with the tracking control off, and to view the rf signal from the heads as it enters the first fm limiter. If the control head is then moved sideways, the level of the rf signal will vary as the rotating heads alter their scan. If the control head is adjusted so that maximum rf signal is observed, the heads will be scanning the recorded tracks correctly.

AMPEX 7500

A simplified block diagram of the servo in the Ampex 7500 is shown in Fig. 8-28. In this machine, a head-tach pulse is produced by a rotating magnet mounted beneath the head-drum assembly; the magnet passes over a small coil once each revolution, and thus a pulse is produced every 1/60 second. The pulse is shaped by two monostable multivibrators to produce a symmetrical square wave, which is integrated by a linearly charged capacitor to produce a ramp waveform.

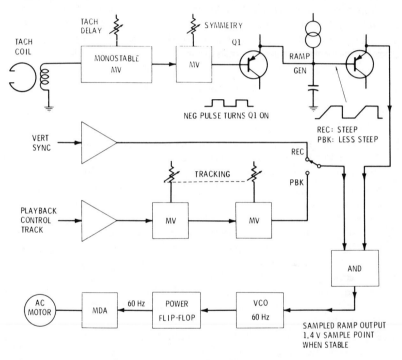

Fig. 8-28. Block diagram of servo in Ampex 7500.

In the record mode, the vertical-sync reference signal and the ramp are fed to an AND-gate comparator. When the servo is operating correctly, the ramp is sampled at about 1.4 volts above ground.

On playback, the control pulse from the tape is the reference. This has 7 to 25 milliseconds of adjustable delay provided by two flip-flops and the tracking controls. The delayed pulse is applied to the AND gate in place of the vertical sync. (This machine is one of the few cases in which the reference signal is used as the sampling pulse instead of being formed into a constant ramp.)

165

The output of the gate is applied to a voltage-controlled oscillator that is set to free run at exactly 60 Hz. This is to permit the motor to run correctly for a short time if the input signals are lost due to bad tape or other causes. The oscillator can be pulled both high and low in frequency by the input, thus allowing the motor to be speeded up or slowed down. Ideally, the output of the oscillator is a 60-Hz square wave that is amplified and then fed to the motor drive amplifier.

The motor drive amplifier consists of four SCRs connected in a bridge inverter configuration that controls the motor speed by reversing the polarity of a 140-volt dc supply. The rate of the reversal is controlled by the frequency of the oscillator. Note that the motor is an ac motor, and on some models the vco is provided with adjustments for both 50- and 60-Hz operation.

SONY EV 320 SERIES

In the Sony EV 320 series, the reference signal for the head servo is either the vertical sync from the video or a pulse derived from the 60-Hz ac line. The block diagram of Fig. 8-29 should help in understanding the following explanation.

The video signal is passed through a low-pass filter to remove noise and reduce the high-frequency video components, and then it is passed through a sync stripper. The output is passed through an integrator to remove the horizontal sync, and the remaining vertical sync is shaped before being used to trigger a multivibrator. The output of the sync separator is also fed to a detector that is used to sense the presence of incoming sync. If sync is not present at the input, a gate circuit allows a pulse from the 60-Hz power line to pass to the input of the mutlivibrator to act as an alternative reference trigger. In this way, the head servo has a steady reference signal at all times.

On record, when an input video signal is present, this is used at all times as the reference. In the playback mode, the power-line reference is used to synchronize the head if no video input is present. But in an editing situation, where external video is supplied, the external video will always be used so that the two video signals will align perfectly. This is obviously required when the change from recorded to incoming video and vice versa is effected.

One output of the sync-separator multivibrator is buffered and fed to the control head in the record mode. The other output is integrated to form a ramp, and the ramp is sampled by the head-tach pulse.

The head-tach pulse occurs at a 30-Hz rate, and it is shaped by a monostable. It is the output of this monostable that is used to sample

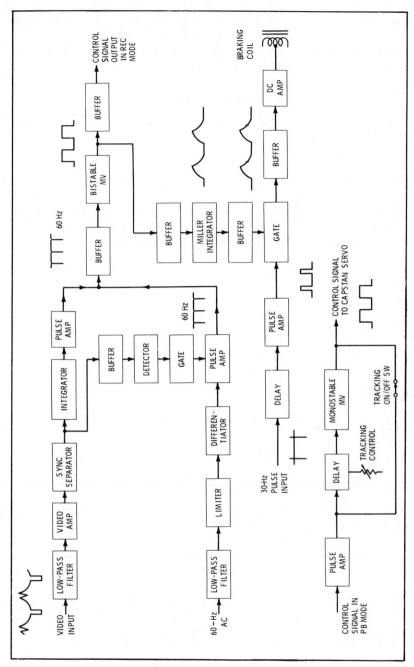

Fig. 8-29. Block diagram of Sony EV 320 servo.

167

the ramp. The sampled ramp is integrated to a dc level to control the current through an eddy-current braking coil.

Note that in the playback mode the control track is applied to the capstan servo and not the head servo.

SONY AV 3600

Fig. 8-30 is a simplified block diagram of the servo in the Sony AV 3600. The incoming vertical-sync pulses are used as the reference, and every other pulse is compared to a 30-Hz tach pulse from the rotating heads.

In the record mode, the separated sync is fed to monostable MV 402. The on time of this stage is set to be 24 ms so that it ignores alternate incoming vertical-sync pulses. The output is fed to monostable MV 404, which is set to give a 30-Hz square-wave output with a 50-percent duty cycle. This is integrated into a ramp and fed to a comparator.

In the playback mode, the amplified pulses from the control track are fed into MV 402, and the output of this stage is fed to MV 404 either directly or through monostable MM 403. The duty cycle of MM 403 is set by the tracking control, which provides a continuous

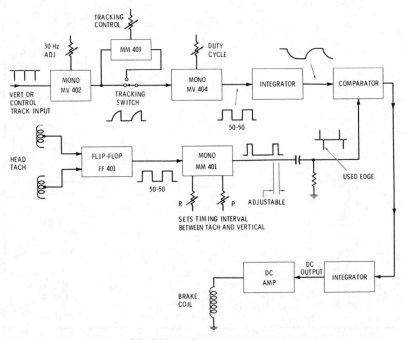

Fig. 8-30. Simplified block diagram of Sony AV 3600 servo.

variation over a 12-ms range. The output of MV 404 is again integrated to a ramp, but the ramp has a longer slope than that used in record.

In both modes, the head-tach pulses are fed continuously to each side of a flip-flop, which is triggered by the alternate pulses. The output from this flip-flop is a square wave with a 50-percent duty cycle; the square wave is used to trigger a further monostable. The output of this monostable is a pulse the width of which is set by a potentiometer control. Separate controls are used for record and playback to produce a different pulse width for each mode. This pulse is differentiated, and the negative-going edge is fed to the comparator. These negative edges are thus delayed by a given time from the input pulse. The different delays on record and playback are used so that the different ramp slopes are sampled at the correct point to produce the same correct dc level. The dc level of the sampled ramp is integrated to a dc level to control the eddy-current braking coil. The differing slopes on record and playback are necessary to bring about the required different rates of servo reaction.

SHIBADEN SV 700

In the Shibaden SV 700 (Fig. 8-31), the incoming vertical sync is used as the reference signal, and the head disc is phased with it to produce correct recording and playback. Two heads are mounted on the rotating head disc, and at right angles to the heads are mounted two small bar magnets. These produce pulses from a single coil mounted above the disc. The head-tach pulses are integrated to a ramp, which is sampled in a diode bridge by a pulse formed from the incoming vertical sync. The pulses at the output of the diode comparator are amplified and then integrated to a dc level. This dc level is used to control the frequency of a blocking oscillator, and the pulses from the blocking oscillator trigger a flip-flop to produce a 60-Hz square wave. The square wave is transformer coupled to the motor drive amplifier (MDA), which directly drives the head motor.

The head-tach pulses, instead of pulses formed from the incoming vertical sync, are recorded onto the tape to provide the control track. After amplification, the head-tach pulses are fed to the video input circuit as a blanking pulse. If the servo is correctly locked up, this produces a black bar in the back porch of vertical sync. If the servo is not locked up on either record or playback, this is immediately obvious because a black bar is seen to run through the picture. This is the only machine with a visual indication of servo lock-up in the record mode.

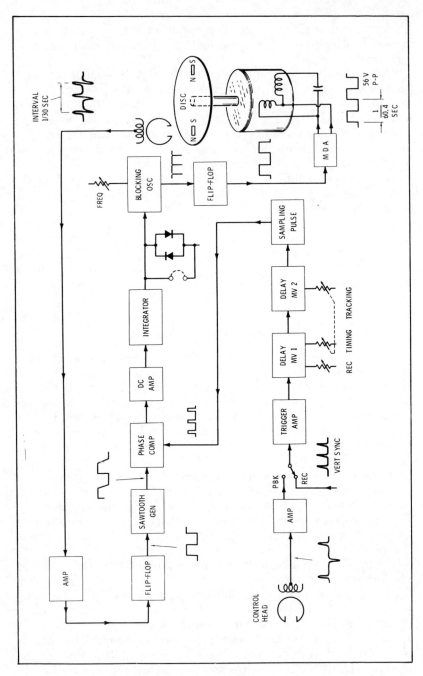

Fig. 8-31. Block diagram of Shibaden SV 700 head servo.

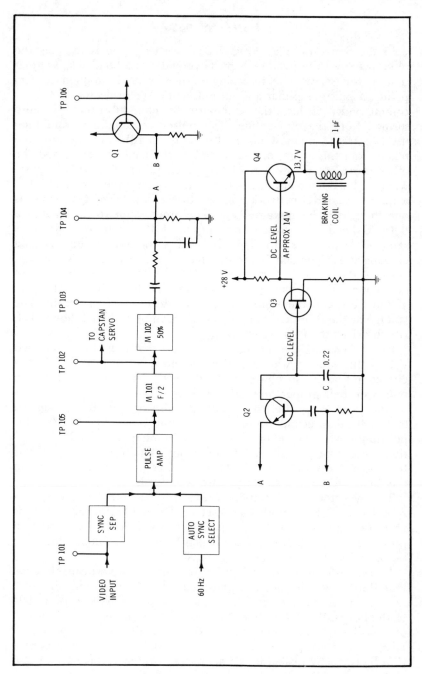

Fig. 8-32. Diagram of Sony AV 3650 head servo.

SONY AV 3650

In the Sony AV 3650 (Fig. 8-32), vertical sync is used as the reference, and the head-tach pulse provides the feedback information to the servo. The incoming vertical sync is applied to two points, a pulse amplifier and an automatic sync selector. If the incoming video is lost, the automatic selector substitutes a pulse formed from the 60-Hz line. The pulses are amplified and then used to trigger monostable multivibrator M101. Two outputs are taken from this stage; one is routed to the capstan servo, and the other provides an input to M102, which provides a 30-Hz square wave with a 50-percent duty cycle. This square wave is integrated to a ramp by an RC network and fed to the emitter of Q2, which is one input of a comparator gate. The other input to the gate is a delayed head-tach pulse. The output of this gate is a dc level sampled from the ramp. This dc level is formed by the charging of capacitor C, which holds its charge after the gate is shut off because the following transistor, FET Q3, has a very high input impedance. The voltage across the capacitor is applied to the gate of the FET and therefore controls the dc level at the drain. The voltage at the drain controls the conduction of emitter follower Q4, which drives the braking coil.

Fig. 8-33 shows the timing relationships that must be maintained within the circuit. Adjustment potentiometers are provided in the circuit so that the machine can be correctly aligned to produce the pulses in their proper time relationships.

Note that one pulse, the sampling pulse on TP 104, is delayed only 900 microseconds from the preceding pulse on TP 103. This position must be maintained to within 10 microseconds. This variation corresponds to the width of a horizontal-sync pulse, or less than $\frac{1}{5}$ of the visible line. If the vertical can be maintained within this time variation, a stable picture will result.

The waveform at TP 106 includes a pulse that occurs 8 horizontal lines (8H) before the onset of the vertical interval. This is used for head switching, and it is derived by delaying the head-tach pulse by using a multivibrator. It can be set by a potentiometer to be placed within about one line of video.

It is in order to produce a pulse in front of the vertical reference that the head-tach coils are placed on the drum roughly halfway around the active sweep of the heads. With the metal vane at 90° to the heads, this arrangement can produce a pulse a short amount of time before the occurrence of the vertical interval. The multivibrators can then be used to delay the pulse slightly and permit it to be correctly placed for comparison with the vertical-sync pulse and to effect head switching.

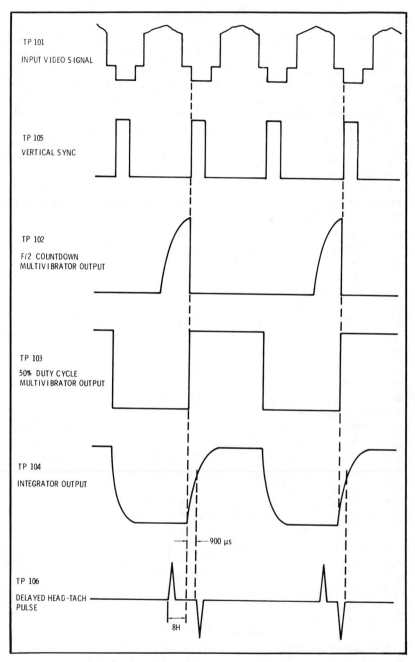

Fig. 8-33. Waveforms in Sony AV 3650 head servo.

The Capstan Servo

In audio recorders and the simplest vtr's, the capstan is driven at the same constant speed on both record and playback. Usually, a synchronous motor driven from the ac power source is used to provide this constant speed under normal operating conditions. In portable machines, a dc motor is used, with a constant dc supplied to keep the speed constant. Both direct and belt drives are used.

When editing is required on a vtr, the situation changes radically with respect to the capstan drive. It is now necessary that the previously recorded video on the tape be aligned accurately with the new incoming video. To do this, the speed of the tape must be varied until the vertical-sync pulses of both signals align, and then the tape must be held at a constant speed to maintain this alignment. To achieve this, the capstan must be servo controlled.

For this purpose, a synchronous motor powered from the ac line is unsuitable. It must be run from a power amplifier that is driven by a controlled internal oscillator, or a dc motor must be used. In either case, the frequency of the oscillator or the dc level must be controlled by an error signal developed from comparison of a reference source and a feedback signal dependent on the linear speed of the tape and its position with respect to the incoming vertical sync.

Another way of doing the job is to use a second, or subsidiary, motor which, like the main motor, belt drives the capstan. Here, the main motor continues to receive a constant drive, and the drive to the subsidiary motor is varied so that it either speeds or slows the capstan rotation.

The main object to be achieved is to control the tape longitudinal motion to allow the rotating heads to follow the center of the recorded tracks precisely and to frame the tracks exactly in step with the incoming video. The following examples show how this is achieved in the machines described. The details of track and video alignment, as well as the operational uses, are left until the chapter on editing. It should be fully understood that the head servo continues to operate in its normal manner in conjunction with the capstan servo. This is necessary because the capstan servo affects the tape linear speed and position only. In machines where a capstan servo is used, the tracking control is usually moved from the head servo to the capstan servo.

SONY EV 320F

The Sony EV 320F provides an example of a very simple capstan servo. During playback, this servo (Figs. 8-34 and 8-35) comes

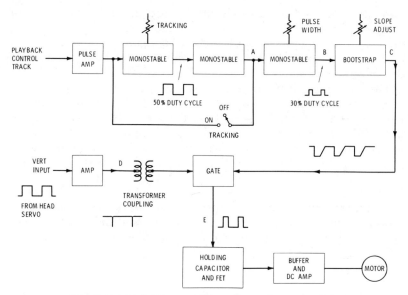

Fig. 8-34. Block diagram of Sony EV 320F capstan servo.

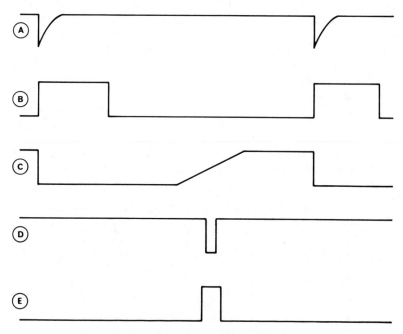

Fig. 8-35. Waveforms in Sony EV 320F capstan servo.

into operation, and the capstan rotation is controlled by a dc motor that belt drives the capstan so as to either aid or oppose the belt drive from the main transport motor. The main motor continues to receive its constant drive, while the servo circuit operates only on the subsidiary motor.

The control-track pulses are referenced against the vertical sync of the incoming new video. If playback only is required, with no incoming external video for editing, then the reference is provided by a pulse derived from the 60-Hz power input. The servo circuit uses a simple ramp-sampling method that produces a dc output to control the speed of the subsidiary dc motor. Note that no feedback is taken directly from the motor. In other words, this is an "open loop" servo.

In the record mode, the capstan is not servo controlled. The power to the subsidiary motor is interrupted, and the capstan motor is driven by preset power to produce a constant tape speed.

SONY AV 3650

An example of a more complicated servo is provided by the Sony AV 3650. The capstan is controlled by both a speed and phase servo. Separate circuits are used, and their outputs are combined to control a dc motor that belt drives the capstan. Both work simultaneously to correct and control the tape speed and position. The dc motor is specially constructed with a tonewheel and pickup coil mounted inside. When the motor is running at the correct speed, the output from the coil is a sine wave of 1020 Hz in older models, and about 965 Hz in later machines. This output is used to produce a ramp, which is then sliced by a preset dc level to give speed control in both recording and playback. For phase control in recording, it is counted down to 30 Hz. In the playback mode, the phase servo uses the control pulse.

Table 8-1 shows the signals that are used as reference and feedback in each servo circuit in both modes. Fig. 8-36 is a block diagram of the circuit.

Table 8-1. Signals in Sony AV 3650 Capstan Servo

	Record			Playback		
	FB	Ref	Function	FB	Ref	Function
Phase	FG Coil	Vert Drive	Capstan Speed Control	Control Track	30 Hz Head Tach	Tracking
Speed	FG Coil	DC Level	Capstan Speed Control	FG Coil	DC Level	Capstan Speed

The service manual for this machine breaks the servo down into three stages: (1) the speed servo, (2) the phase servo in the recording mode, and (3) the phase servo in the playback mode. In

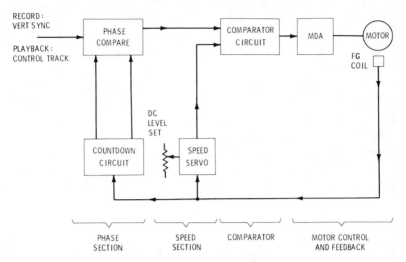

Fig. 8-36. Block diagram of Sony AV 3650 capstan servo.

this description, the comparator and MDA are treated separately at the end.

Speed Control Servo

The operation of the speed control part of the servo circuit is the same on record and playback. A ramp-slice comparator is used in both modes.

A sine wave is produced by the frequency generator (FG) coil in the motor and is amplified to a square wave in the limiter section (Fig. 8-37). The square wave is differentiated to a series of pulses that are used to reset a continuously running sawtooth generator. The sawtooth waveform is the input for one side of a differential amplifier; the other side is held at a constant dc level set by a potentiometer. The output of this stage is a series of pulses the width and spacing of which depend on the dc level and the input frequency. From the sine-wave input to this point, everything is accomplished in one integrated circuit.

At this point, the pulses could be integrated to drive a dc motor, but instead they are fed to a comparator where they are used as a reference working against a varying dc input from the phase servo. The comparator uses two transistors in the IC; the rest of the circuit is made up of discrete components.

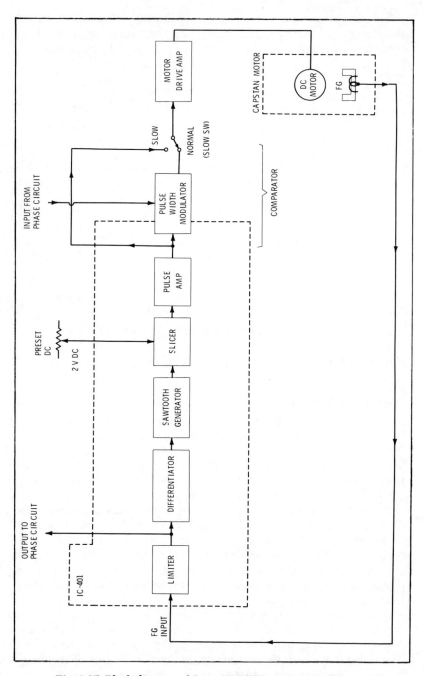

Fig. 8-37. Block diagram of Sony AV 3650 capstan speed servo.

The overall action is that variation in the motor speed varies the output frequency from the coil. Comparison of this output to a steady dc produces a voltage to correct the motor speed.

Capstan Phase Servo Control

The phase servo control uses a basic ramp-sampling method with a dc output. The dc output is applied to the comparator section and is used to refine the capstan control. In the record mode, it corrects for minor drifts from the incoming vertical sync, and in playback it maintains alignment with the rotating heads to assist in correct tracking.

The circuit action is quite different in each mode, and the modes are treated separately. Reference to Table 8-1 shows that both inputs change when a change of mode is selected.

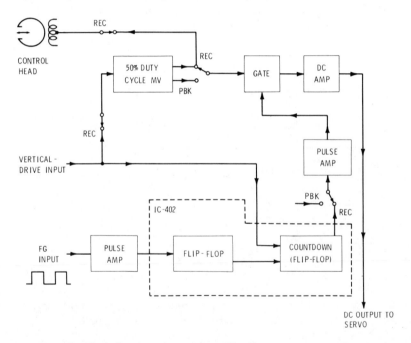

Fig. 8-38. Block diagram of Sony AV 3650 phase servo in record mode.

Record Mode—Fig. 8-38 is a block diagram of the record process for this servo. Vertical sync is used as the reference and is phase compared to a specially derived internal signal. This signal is formed from the counted-down FG signal, which has also been compared to the incoming vertical. The FG square wave from the speed-servo limiter is divided and compared to a vertical-drive input from the

179

head servo circuit. This produces an output pulse as at D in Fig. 8-39B.

This is a pulse train with a repetition rate of 30 Hz. The position of the negative-going edge depends on the FG frequency, and it moves as that frequency is varied. When this waveform is differentiated to a series of spikes, the negative spikes are used as sampling pulses for a ramp. On record, this ramp is derived from the

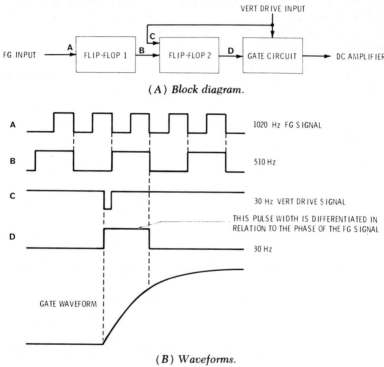

(A) Block diagram.

(B) Waveforms.

Fig. 8-39. Tonewheel pulse circuit and waveforms.

incoming vertical sync, and it is sampled on its leading slope to produce a dc level. This dc output serves as an input to the phase-speed comparator.

Playback Mode—Fig. 8-40 is a block diagram of the playback mode of operation. This circuit is unusual in its choice of inputs and controls. The reference pulse is a 30-Hz square wave from the head-tach coils instead of the vertical sync, and the feedback is from the tape control track.

Fig. 8-41 shows a simplified schematic diagram of the gate and sampling section of this servo. The control-track pulses are amplified and then converted to a square wave, which is integrated to a

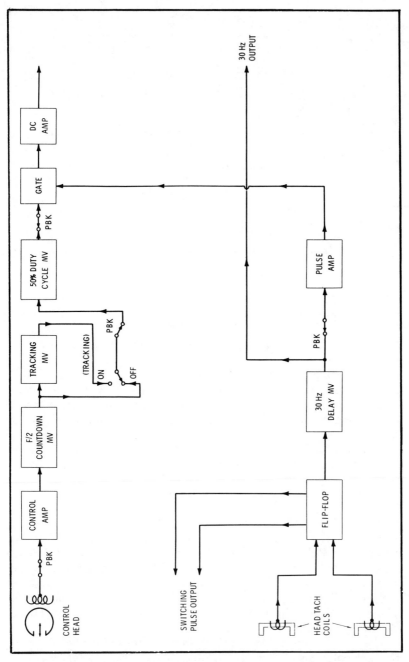

Fig. 8-40. Block diagram of Sony AV 3650 phase servo in playback mode.

181

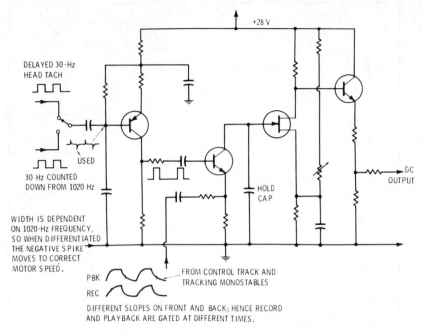

Fig. 8-41. Simplified circuit diagram of Sony AV 3650 phase servo.

ramp for sampling; the longer slope of the negative-going trailing edge is used. The onset of the negative edge of the square wave can be varied by the use of the tracking control whenever necessary. This is one of the few cases where the tracking control is used in the capstan servo and not in the head servo.

The head-tach pulses produce a 30-Hz square wave with a 50-percent duty cycle. This is differentiated to produce a series of pulses, and the negative spikes are used to effect sampling. Again, the final output is a dc level to the servo comparator.

Comparator and MDA

Fig. 8-42 is a basic schematic of the comparator and MDA. The outputs from the two servos are compared in a ramp-slice comparator circuit, which is used to control the motor drive amplifier. The output from the speed servo is a series of ramps, as shown in Fig. 8-43. These are sliced by the varying dc from the phase servo. The output is a train of pulses the width and spacing of which can be varied by either the speed or phase servo.

The input to the MDA is this train of pulses, which is integrated by the motor and other circuit elements in the emitter circuit of the final transistor to produce a stable drive to the motor.

182

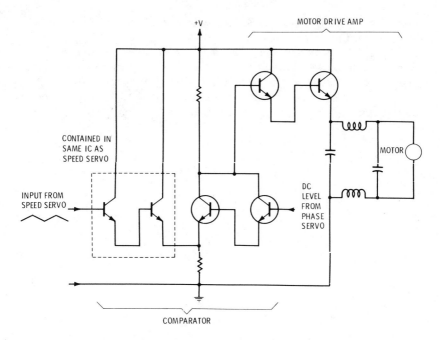

Fig. 8-42. Comparator and MDA in Sony AV 3650 capstan servo.

Fig. 8-43. Ramp slopes.

Conclusion

From the preceding sections, it can be seen that many different methods of servo control have been used, with varying degrees of success. The choice of method seems to have been made by individual designers who worked on the model in question, which might account for the variation of methods and of circuits within the same method.

For head control, the sophisticated eddy-current braking method seems to be the most popular, since it is easy to achieve and is effective. Most of the interchange and tracking problems in vtr's are due to drift of the servo out of alignment, and all manufacturers' service manuals cover servo lineup and troubleshooting quite fully.

The choice of direct or belt drive is not an easy decision, and differing opinions do exist. The choice of dc or ac motors also is

made on an individual basis, and it is a decision not entirely separate from the belt controversy.

The purpose of this chapter has been to give an idea of basic
servo action and to present a few examples of those in common vtr
use. It is by no means a complete or exhaustive treatise, but it is
enough to enable the reader to follow any vtr servo in principle
and practice.

9

Quad Servos

In this chapter, the basic circuits of the head and capstan servos used in quad machines will be described briefly. The quad servos differ in some respects from those normally found in helical machines. The actual circuits are complex, and full descriptions would require so much space that the details are best studied by consulting the service manuals for the particular machine. Although individual models have different circuits, they are all based on the simple principles outlined here.

One important difference between the heads in the helical and quad machines is the much higher rotational speed of the quad head drum, and the fact that it must be much more tightly controlled in both speed and phase. The head motor is usually driven by a 240-Hz signal, and feedback is taken from the motor by magnetic or optical means.

The tape in most quad machines travels at 15 in/s, whereas tape speeds in helical machines range from about 3¾ to about 9½ in/s. Because of the transverse tracks on the tape in the quad format, the quad capstan must be much more accurately controlled than the helical capstan. In most cases, the capstan motor is driven by a 60-Hz signal, but usually no feedback is taken directly from the motor.

Two types of capstan and head motors are used in the quad machines.

1. The motor is treated as a synchronous-induction motor. The excitation frequency is much higher than the desired operating speed, and speed control is achieved by amplitude modulation of the power.

2. A synchronous motor is used. A combination of frequency and phase modulation is applied to the drive signal to maintain control.

Motor-drive methods are covered in the next section before the actual servo circuits are discussed. The reason for this is that the motor-drive method dictates the final part of the servo circuitry, which is quite different from that seen in the small helical machines.

The reference signals for both the head and capstan servos are obtained from either the vertical-sync pulse of the incoming video, an external sync generator, or the 60-Hz power line. In the record mode, the reference is converted up to a 240-Hz sine wave and recorded along the bottom edge of the tape by a conventional static head. This sine wave is used as a reference signal in the playback mode. Fig. 9-1 shows the phase relationship of the sine wave to the video tracks. This ensures correct head tracking in playback. The position of the 30-Hz edit pulse permits all edits to be made in the vertical interval.

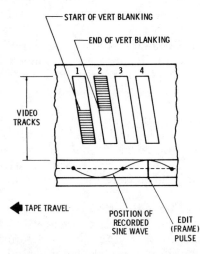

Fig. 9-1. Phasing of control track relative to video tracks.

The feedback information for the servos is provided by the head-tach pulses. The position of the pulse is important, because it indicates the position of the head that records and plays back the vertical interval.

The head-tach pulses may be generated by a magnetic method or a photoelectric method. In the photoelectric method, a wheel, painted half white and half black, is mounted on the rotating head shaft (Fig. 9-2). Light from a lamp reflects from the white section into a phototransistor or photocell, but the black section produces no reflection. The output of the photosensor is thus a square wave.

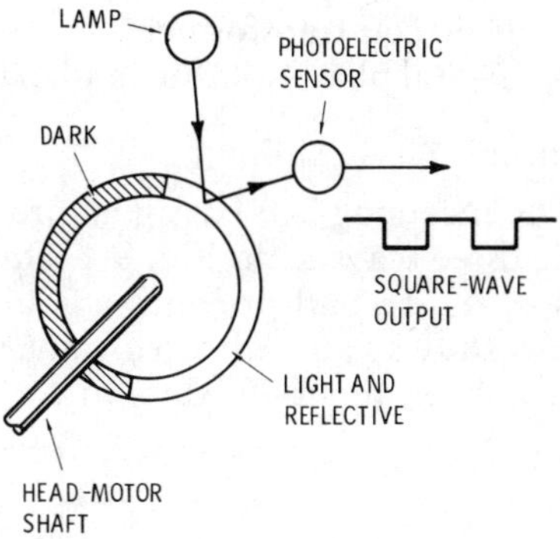

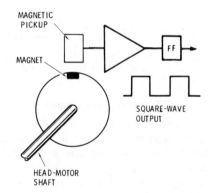

Fig. 9-2. Photoelectric method of generating head-tach pulses.

The frequency of the square wave is indicative of the head speed, and the transitions indicate the head position. When the head is rotating correctly, a 240-Hz square wave is produced.

Fig. 9-3. Magnetic method of generating head-tach pulses.

In the magnetic method (Fig. 9-3), a small magnet is embedded in the circumference of the wheel, or the wheel has a small notch in it. Either of these will produce a pulse from a coil that is placed close to the wheel. Again, the proper output frequency is 240 Hz.

HEAD AND CAPSTAN MOTOR DRIVES

Before the simplified head and capstan servos are described, the methods of driving the motors should be covered. In quad machines, belt drives and magnetic brakes are not used. Three-phase motors that require a 240-Hz drive signal are used. These provide a very smooth drive, and fine control is easy to accomplish with them. Three basic methods are used to provide the three-phase drive to the motors:

1. Electronic phase shifting
2. The Scott transformer
3. Digital phase splitting and drive

Electronic Phase Shifting

The incoming 240-Hz signal produced by the servo circuits can be split three ways as in Fig. 9-4. One path provides a 120° phase advance, the second provides a 120° phase lag, and the third has no phase shift at all. All three signals are applied to power amplifiers that deliver about 70 watts to each of the motor windings.

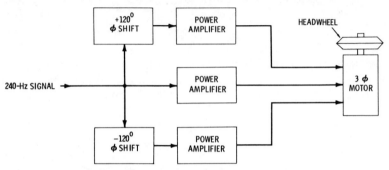

Fig. 9-4. Motor drive with electronic phase shifting.

The Scott Transformer Connection

In another method, the output from the servo circuit is both advanced and retarded by 45°, and the two resulting signals are amplified. They are then applied to different primaries of a Scott-connected transformer assembly (Fig. 9-5). The two secondaries produce a three-phase output that is applied to the motor windings.

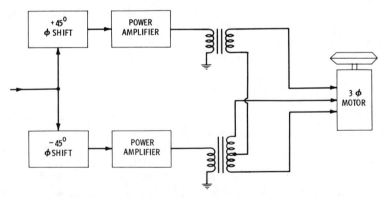

Fig. 9-5. Motor drive with Scott-connected transformers.

Digital Drive

The three phases can be formed by a digital circuit. This is a more recent approach than the other methods and is finding much use in the later machines. It is easy to generate and control the digital pulses, and the circuits are much easier to service. A digital three-phase generator is shown in Fig. 9-6. The input signal is from the output of the servo circuit, and the three phases are applied to the motor windings.

Fig. 9-7 shows a digital three-phase generator placed in the early stages of a servo circuit. An 8H signal is developed from the incoming sync and is applied to the three-phase generator. The three phases are applied to identical modulator circuits, where the rising and falling edges of the three square waves are position modulated by the error signal from the servo comparator. These output square

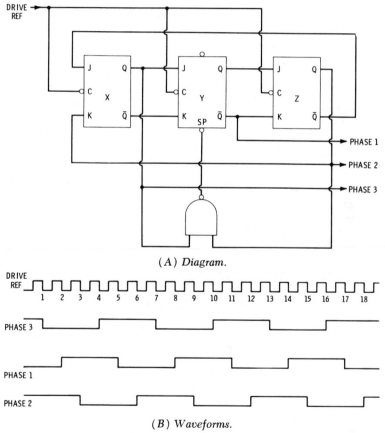

(A) Diagram.

(B) Waveforms.

Fig. 9-6. Three-phase ring counter.

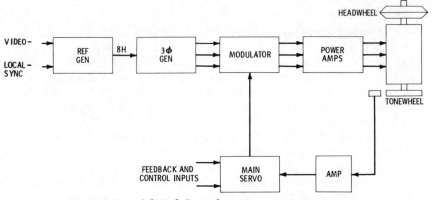

Fig. 9-7. Use of digital three-phase generator in servo system.

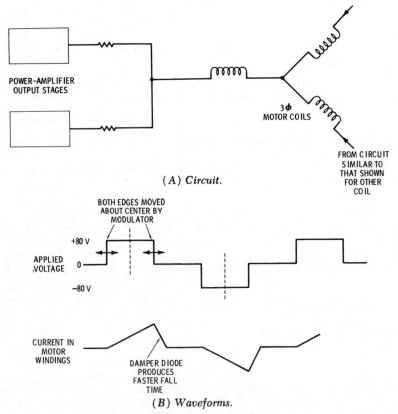

(A) Circuit.

(B) Waveforms.

Fig. 9-8. Three-phase motor drive.

waves are fed to the power amplifiers and then to the motor windings.

The motor is used as an induction motor, and at the operating frequency the windings act as inductors so that a fairly linear current build-up occurs. Thus the voltage square waves produce triangular current waveforms in the motor windings to produce the drive. Fig. 9-8 shows the waveforms of the width-modulated voltage pulses, which, in effect, produce amplitude-modulated current pulses.

THE HEAD SERVO

A block diagram of a head servo is shown in Fig. 9-9. This circuit is easiest to understand in the record mode, so this mode will be covered first; the playback mode is covered briefly in a later section. There are two servo loops in this circuit. The first provides a fast phase correction, and the second minimizes hunting.

The motor is driven by a 240-Hz signal that is derived from one of the three main 60-Hz reference sources. In the first loop, this 60 Hz is multiplied to 240 Hz, split into two phases, and passed into a comparator. It is compared here with two phases obtained from the 240-Hz tonewheel signal. The comparator output controls a 240-Hz oscillator. The output of the oscillator is formed into a sawtooth waveform, which is then clipped to form a trapezoid. The trapezoid is used to form the motor-drive signal. The two inputs from the tonewheel are directly dependent on the position and speed of the heads. Large errors in the sync phase will not produce an immediate effect on the oscillator, but they will cause it to shift its frequency gradually and gradually shift the head phase.

The second loop in this servo circuit uses only the 240-Hz signal from the tonewheel. This signal is split into two paths. The first goes to a ringing oscillator that produces a 240-Hz square wave displaced 90° from the input. This is not an exact 90° difference, since it varies depending on slight variations in the speed of the head. The other path for the signal is through a phase splitter into a phase comparator, where it is compared with the two phases from the oscillator phase splitter. If the head velocity should change slightly, then the comparator output will change because the tonewheel signal will not be exactly at 240 Hz, and thus will differ from the oscillator signal, which is always at 240 Hz. The error voltage from the comparator samples or modulates the sawtooth waveform, to produce a voltage that controls the 240-Hz multivibrator that drives the motor.

Other head-servo circuits exist, but they are all similar to the one described here. They all produce a very tight control of the position and speed of the heads.

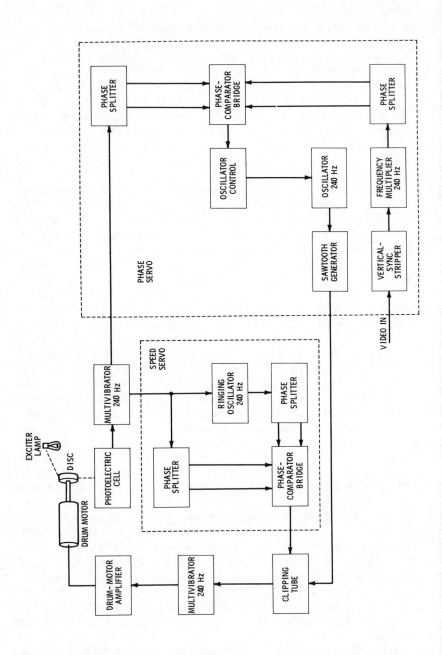

Fig. 9-9. Simplified block diagram of head-drum servo.

THE CAPSTAN SERVO

The capstan servo is slightly different from the head servo; a simplified block diagram is shown in Fig. 9-10. In the record mode, the reference frequency for the capstan is the 240-Hz tonewheel signal. Thus the capstan is locked to the headwheel rotation, and therefore to the reference input that drives the headwheel. The 240-Hz signal from the tonewheel is divided down to 60 Hz and used as the reference for the 240-Hz oscillator that drives the capstan motor. Feedback from the oscillator output is compared with the tonewheel signal to control the oscillator output frequency. The output of the oscillator is amplified and used to drive the motor. Note that no feedback is taken from the motor, so this is an "open loop" servo system.

The tonewheel output is delayed and recorded onto the control track of the tape.

In the playback mode, the motor-drive oscillator is controlled by a comparison of the 240-Hz signals from the control track and the tonewheel. Normally, these will be 90° out of phase, and any error varies the oscillator frequency slightly and corrects the capstan speed.

SERVOS IN THE PLAYBACK MODE

In playback, there are four distinct modes of operation for the head and capstan servos. These modes are necessary in broadcasting and teleproduction, where the output of the vtr must be synchronous and color phased with other vtr's and cameras.

1. The *tonewheel* control mode is the simplest. The capstan moves the tape so that the heads are centered on the recorded tracks, and the headwheel moves at the correct speed to give a stable picture. No phasing between playback sync and local or station sync is used. This method is used for simple playbacks only.
2. In *switchlock*, the capstan controls the tape so that when the vertical interval of local sync occurs, the track containing the recorded vertical sync is being scanned. The headwheel operates just as in the tonewheel mode. This coarse form of vertical framing provides a close line-up of the two vertical intervals, but it does not compensate for any recorded errors in the placing of the vertical sync.
3. In *linelock,* the capstan servo references itself against the headwheel. The headwheel can lock onto any horizontal line, and this allows a quick recovery from tape disturbances such as splices, dropouts, etc.

4. The *pixlock* mode is the tightest mode available in current quad machines. The headwheel is phased so that precise lock-in is obtained between the playback horizontal and vertical sync and the incoming horizontal and vertical sync. The capstan works in the switchlock mode. In this mode, the playback sta-

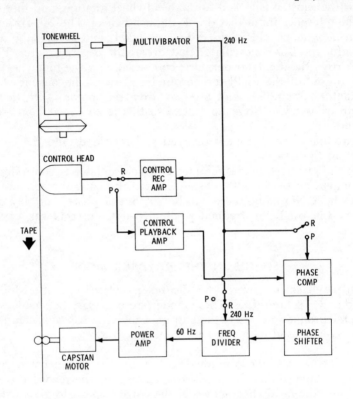

Fig. 9-10. Simplified block diagram of capstan servo.

bility is within about 0.15 microsecond. This tight control is essential so that the analog time-base correctors can produce stable color playback and edits.

All broadcast quad machines have added circuitry for these extra levels of control; the added circuits have not been shown on the simple block diagrams presented here. These tighter modes of control are usually achieved by extra servo loops that use the horizontal sync and the incoming subcarrier as references for comparison with similar signals from the tape playback.

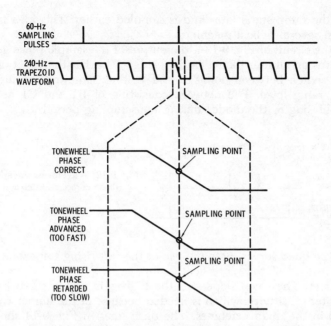

Fig. 9-11. Waveforms in headwheel phase error circuit.

SERVO COMPARATORS

In most of the quad servos, the ramp-sample type of comparator is used in both the phase and velocity (speed) loops. The main difference from one to another is the frequency of the input signals. In all cases, the higher frequency is formed into the ramp, and the lower one is formed into the sample pulse.

In the headwheel phase error circuit, the tonewheel 240-Hz signal forms the ramp, and the 60-Hz reference signal forms the pulse (Fig. 9-11). In the capstan servo, both frequencies may be 60 Hz, but in this case only alternate pulses may be used to sample the ramp. See Fig. 9-12.

The velocity, or speed, error circuit shown in Fig. 9-13 is interesting in that the tonewheel signal is used to form both the ramp and the sample pulse. If the input frequency rises, the second ramp in waveform E in Fig. 9-13B starts earlier than it would if the frequency were correct. But the sampling pulse is formed from a one-shot delay circuit triggered by the incoming signal, and it is timed to occur 1/240 second after the tonewheel pulse, regardless of the tonewheel frequency. Therefore, if the frequency is high, the sample pulse will appear later on the ramp. The resulting dc output thus changes, and it corrects the headwheel speed. If the input frequency falls, the ramp starts later and is sampled earlier. This idea is used also in several helical machines.

In the circuit of Fig. 9-13A, the output of the sample-pulse generator turns on all four diodes in the bridge. This has the effect of closing a switch between points A and B, thus sampling the instantaneous ramp level. The network consisting of R1 and C1 provides back biasing of the diodes during nonsampling periods.

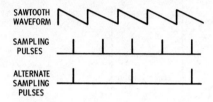

Fig. 9-12. Waveforms in capstan phase comparison circuit.

In the quad servos, note the use of the fast-rising trapezoid waveform instead of the more gently sloping ramps used in the helical machines. The steep slope and the high gain of the servo provide a greater sensitivity, which is needed because of the higher rotating speed in the quad machines. The dash lines in Fig. 9-14 show the relative amplitude of a sawtooth wave with the same slope as the trapezoid wave. Only a small part of the slope is needed, so the trapezoid wave can be used instead of the sawtooth.

SUMMARY

The servos in quad vtr's control the tape and head motion at all times, and a monochrome picture can be produced on playback with timing errors in the neighborhood of 1 microsecond. When these machines first appeared in the mid 1950s, the circuits all used electronic tubes. Later, transistor versions appeared, and after that ICs were gradually introduced. With this reduction in size, the circuits became more complex and were able to exert more exacting control over the servo operation. Eventually, digital control techniques were introduced. These have several advantages:

1. Digital circuits are both more reliable and more accurate than their analog counterparts.
2. They are easier to understand and to service.
3. It is easy to incorporate warning lights to show if the servo is locked and to warn the operator that an adjustment may be required.
4. The indicator lights can also show where a fault may be located, and which signal may be missing.

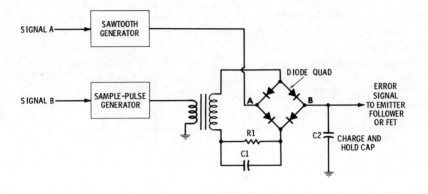

(A) Phase comparator.

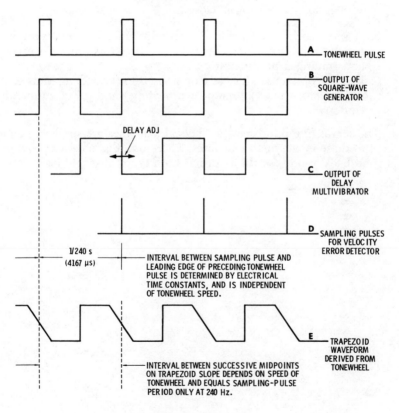

(B) Waveforms.

Fig. 9-13. Velocity error detection.

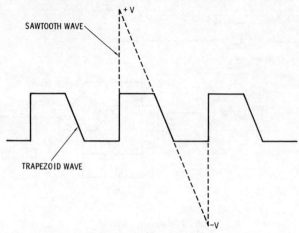

Fig. 9-14. Comparison of trapezoid and sawtooth waveforms.

5. If servo lock is not achieved, an edit function (for example) can be inhibited. This can prevent bad edits from occurring and ruining a production.
6. Further protective devices can be used to show dangerous conditions (for example, motor overheating) and effect automatic shutdown.

The servos found in broadcast helical machines are similar in concept to those in the quad machines. Therefore, this chapter provides a good background for the basics of both types of machines.

10

Control Pulses and
Other Functions

In the process of recording and playing back the video signal, the various pulses and signals produced by the vtr are called upon to perform functions other than controlling the head and tape motion. This chapter describes some of these essential functions and the signals they require. The following is a list of the headings given to these sections, which in general are self-explanatory.

1. Dropout Period
2. Head Switching
3. Dropout Compensation
4. Slow Speed and Still Frame
5. Meters

DROPOUT PERIOD

In a single-head machine, the record-playback video head must cross from one edge of the tape to the other at some point. In both the Ampex and IVC machines, the head begins its scan at the bottom edge of the tape and leaves the tape at the top edge.

At the changeover point, the tape can be arranged to have minimum separation of its edges, or they can even be in contact. However, the changeover does cause some interference in the smooth operation of recording and playback. In Ampex machines, the video tracks do not run from edge to edge of the tape, so a definite time exists when no video is recorded. In the IVC format, the video tracks cover the whole width of the tape, and the dropout time is reduced.

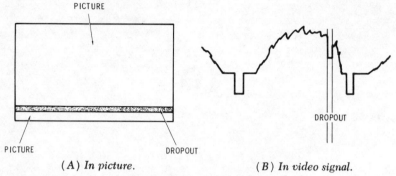

(A) *In picture.*   (B) *In video signal.*

Fig. 10-1. Location of dropout period.

For minimum interference with the video signal, the head position and the changeover point on the tape must be accurately controlled with respect to the video signal. Also, the dropout point must be firmly placed within the video signal, and its signal level and duration must be controlled. Ampex and IVC have adopted somewhat different methods of dealing with this problem.

In the early Ampex machines, the dropout was placed a few lines before the vertical interval (Fig. 10-1), and in the later models it was moved into the vertical blanking. In each case, it lasted about 8–10 lines, and an artificial gray level was inserted at this time. This was generated by the internal dropout compensator, which was also used for tape dropouts. The correct placing of the dropout point occurs when the servos are controlling the heads correctly.

With the alpha wrap and the video tracks extending from edge to edge, the IVC format has a minimal dropout time. When the servo is correctly aligned, the dropout is automatically placed roughly in the middle of the back porch of the vertical-blanking interval (Fig. 10-2).This choice of position is obviously to minimize picture interference, and the dropout time is brief enough to occur here without causing problems with the vertical sync.

If the servo is improperly aligned or has drifted, the dropout can position itself in the vertical sync and cause picture rolling, or it can

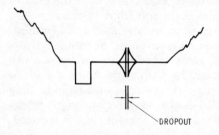

Fig. 10-2. Dropout point in IVC vtr.

appear close to the top of the picture where it will cause a hooking effect of the first few lines.

Adjustment of the tracking control on playback narrows the width of the dropout gap, and correct tracking will reduce the gap to a minimum. Attaching an oscilloscope to the correct test point on the video board or viewing the output on a waveform monitor will show this condition.

HEAD SWITCHING

In a one-head machine, head switching obviously is not required, but in two-head machines, it is necessary to switch from one head to the other in order to have continuous recording and playback. In contrast to the one-head machine, the tape wrap and the heads in a two-head machine can be arranged so that both heads contact the tape for a short period of time, thus completely eliminating the problem of signal dropout due to head changing. The first stage of each preamplifier can be switched on or off by a switching transistor that is controlled by a pulse derived from the head-tach sensors.

The output from the head-tach sensors is fed to opposite sides of a flip-flop, the output of which changes state every time a head-tach pulse occurs. This is the same flip-flop used in the servo circuits, and the outputs from it are fed to the switching transistors as well as the servo. In this way, the preamplifiers are alternately turned on and off, and the switching occurs only at the correct time and when the heads are in the correct place. With electronic switching, the change-over of the heads can be set precisely and can be accomplished in a very few microseconds.

Two positions for the head switching have been used, and they are about the same places used in the one-head machines for the dropout position. If the switching occurs just before the vertical interval, it can be barely discerned in the video. In fact, often the only evidence of its presence is that the first few lines of video after the switch do not align exactly with the previous video. This is due to minute dihedral and other misalignments and causes a very slight flagging at the bottom of the picture. Switching in the back porch of the vertical interval can be seen as noise in the black bar of vertical blanking if the vertical control is adjusted.

Fig. 10-3 shows the relationships of the pulses, head positions, and video signal during head switching. In general, electronic adjustments are provided, and the service manuals have procedures for checking the pulse timing. Ensuring the head-tach sensors are 180° apart is a mechanical problem that is also covered in the manuals, with exhortations to care in the handling of all parts in the head-drum assembly.

201

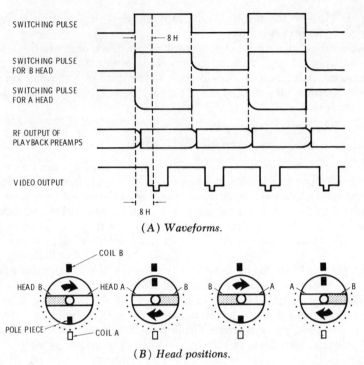

(A) Waveforms.

(B) Head positions.

Fig. 10-3. Head switching.

The circuit of Fig. 10-4 is typical of those found in a modern helical vtr. The FET (Q1) is the first stage of the playback preamplifier, and Q2 is the switching transistor. The input to its base is obtained from the head-tach flip-flop.

Alternatively, the switching transistor can be placed as in Fig. 10-5. It is controlled as before.

DROPOUT COMPENSATION

Dropout compensation has nothing to do with the dropout time previously discussed. The purpose of this facility is to deal with uncontrolled and random loss of video from the tape, generally caused by imperfections in the tape or intermittent dirt and dust between the heads and the tape. Either of these conditions will cause a loss of rf due to separation of the tape and heads or due to loss of coating on the tape. If the loss of rf is sensed before it reaches the video output, it is possible to generate a switching pulse and use this pulse to switch in a delayed video signal for the remainder of the horizontal line.

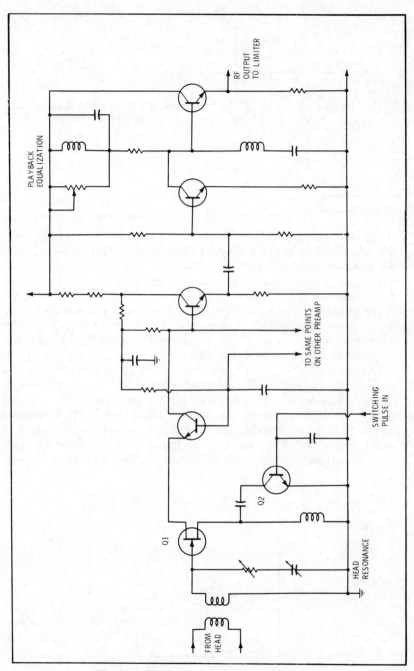

Fig. 10-4. Head switching in preamplifiers.

203

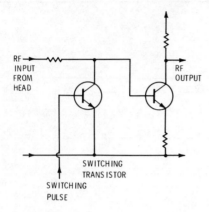

Fig. 10-5. Alternative placement of switching transistor.

Fig. 10-6 is a typical block diagram of the process, and Fig. 10-7 is a simplified circuit of a dropout compensator. The fm signal from the heads is amplified and split into three outputs. The first is fed directly to a dual-input amplifier. The second output is fed into a delay line, and the output of the delay line is fed to the other input of the amplifier. The third signal is rectified, and the resulting dc is fed to a switching circuit. The output of this circuit is governed by the presence or lack of an fm signal, and it is used to switch the amplifier from one input to the other.

The normal state is to allow the direct signal to pass through the amplifier to the output. If a dropout of fm signal occurs, the switch changes state and immediately changes the amplifier to the delayed signal. The delay line is set to retard the signal exactly one horizontal line in time, so on switching over, a section of the previous line of video is repeated. Since most of the imperfections are considerably less than a line in length, this suffices to produce an acceptable sig-

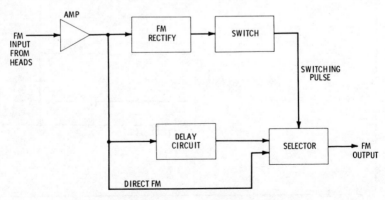

Fig. 10-6. Block diagram of dropout compensator.

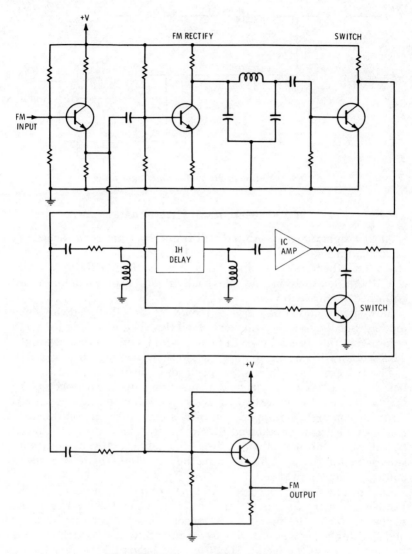

Fig. 10-7. Circuit of dropout compensator.

nal with no flashes on the screen. At the end of each horizontal line, the amplifier switches back to the direct signal.

In Fig. 10-8, a minor variation of the arrangement is shown. In this case, the second input to the amplifier is actually a delay of its own output instead of a delay of the input.

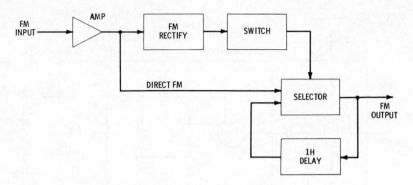

Fig. 10-8. Variation of dropout-compensator design.

SLOW SPEED AND STILL FRAME

At first sight, slow speed and still frame may not seem necessary on a small machine, but for sports instruction, quality control, work study, etc., they are invaluable. Since it is relatively easy to incorporate these features in the playback mode, many machines are equipped in this way.

Slow-speed and single-frame recording are included in some models, but they are not universal facilities. The main use for them is in monitoring breakdowns in factories and similar random events. This section will cover only playback in these two modes.

The principle on which slow-speed and still frame operation are based is to play back one field many times over. To do this, the heads continue to run at their normal speed, but the tape is slowed down or completely stopped. The heads can be servo controlled as usual, or they can be allowed to run freely. The capstan is slowed by adjusting the normal drive from the servo, or by using a subsidiary motor drive. Whichever method is used, there is a speed control on the front panel.

Due to the slower speeds of the tape around the head drum, the heads no longer accurately scan the record video tracks. This is not serious as far as the fm signal is concerned, since the amplification and limiting process will take care of the sections in which there is low output from the heads. However, the heads take longer to traverse the length of the tracks, so the vertical intervals are spaced farther apart. Also, the tape speed is often not constant, so the time between the vertical intervals is varying. This causes a variation in the triggering of the monitor vertical stage, and thus the picture tends to bounce and is generally unstable. To prevent this condition, an artificial vertical pulse is inserted into the video signal just before the final output stage.

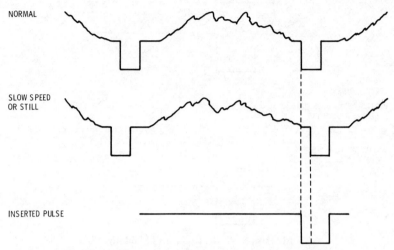

NORMAL

SLOW SPEED
OR STILL

INSERTED PULSE

Fig. 10-9. Relationship of inserted pulse to vertical sync.

The pulse is of sufficient amplitude and duration to mask completely and blank out the unwanted vertical-sync interval already contained in the video signal, and it will trigger a monitor as a normal vertical-interval pulse would. Fig. 10-9 shows the relationship of these pulses in the video signal.

It should be understood that the linear speed of the tape does affect the actual position of the playback vertical interval. The maximum displacement that can occur is in the still-frame mode. The recorded tracks are on the tape as depicted by line AB in Fig. 10-10. In the still mode, the head actually traverses path AC. In a slow-speed condition, the head ends its scan somewhere between points B and C. The maximum time of late arrival of the vertical interval is given by the time it takes the head to scan the section of the track marked x on path AB. Due to the very small angle θ, x is almost equal to distance D. This is about 0.01 of the total track length, so

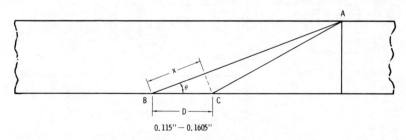

Fig. 10-10. Head paths across tape.

207

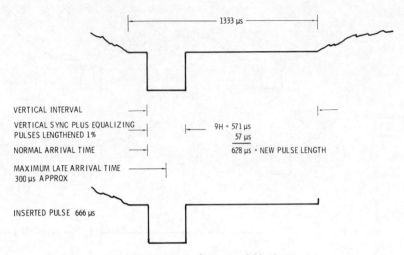

Fig. 10-11. Vertical sync and blanking.

the maximum extension of the time it takes to play back one field is approximately 1 percent, which is less than 300 microseconds. Fig. 10-11 shows the vertical sync and blanking interval of a tv waveform, with the position of a maximally late sync pulse indicated. The bottom waveform in Fig. 10-11 shows the effect of inserting an artificial pulse at the correct time into a lengthened field. Since Fig. 10-11 shows the worst-case conditions, it is obvious that the problem of vertical instability can easily be overcome. (It should be noted that the actual time-extension figures vary from machine to machine.)

Because the heads are scanning under the control of the servo, the head-tach pulses occur at a 60-Hz rate and can be used to produce

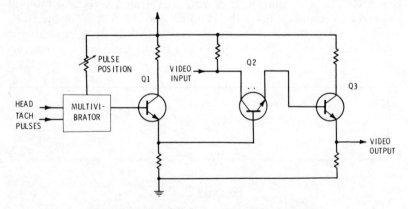

Fig. 10-12. Simplified circuit for pulse insertion.

the inserted pulse. An uncontrolled head runs approximately 1 percent fast, but this is still good enough since a 1-percent difference will not cause monitor problems.

In a slow-scan situation, the heads effectively scan the same path across the tape many times, although it is changing very slowly. Due to the ratio of the track width to the guard-band width, there are times when the head actually will not scan a section of tape that contains the vertical interval. This occurs often enough to make the artificial insertion necessary.

Fig. 10-12 is a simplified circuit of one method of inserting the pulse. Normally Q1 is held on, which holds Q2 on and allows the video signal to pass through to Q3 and then to the output of the machine. When the slow mode is initiated, Q1 becomes controlled by the multivibrator and is turned off for short periods at a 60-Hz rate. These periods correspond to the vertical-sync pulses. These now turn Q2 off, thus inserting an effective vertical-sync pulse into the video signal.

METERS

There are three types of meters in use on a helical vtr, and they are used for entirely separate functions. These are the video meter, tracking meter, and audio meter. When an audio meter is used, it is no different from a meter found in an audio recorder, and thus it will not be covered in this discussion.

Video Meter

The video meter is used to monitor the video level on both record and playback. It is set to give a steady reading that is indicative of the video level. A video level meter does not vary like an audio

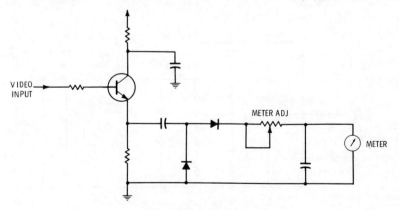

Fig. 10-13. Circuit for video meter.

meter, and there are two main reasons for this. First, the average video level does not change as rapidly as the level of an audio signal, and, second, the changes that do occur in a video signal are much too fast for a meter to follow.

The video signal is picked off at some convenient point and then fed to a buffer circuit. The output of the buffer is rectified to a dc level, which is determined by the average content of the video signal. The dc level is then applied directly to the meter. The meter is calibrated to give a certain reading for a given input signal level. Therefore, a low-level picture will give a lower reading, and an overmodulated signal will give a higher reading. Fig. 10-13 shows a simple practical circuit of such a meter.

This method of video indication is not as good as a waveform monitor, but it is definitely better than nothing. It is certainly good enough for nonbroadcast applications, and it can be most useful to the amateur who does not understand the electronics of television.

Tracking Meter

The object of a tracking meter is twofold: (1) to set the tracking control so that the heads are scanning the tape correctly, and (2) to indicate bad tracking before it appears on the screen.

When the heads are following the recorded tracks correctly, there is maximum output from the heads. Any mistracking causes a reduction in the level played back from the heads. If a sample of the playback signal is taken at a point ahead of the limiter, then the level of this sample will be an indication of the tracking. The sample is amplified, buffered, and then rectified to dc, which is used to drive a meter. Fig. 10-14 shows a typical circuit. Correct tracking gives a maximum reading on the meter; incorrect tracking gives a lower reading.

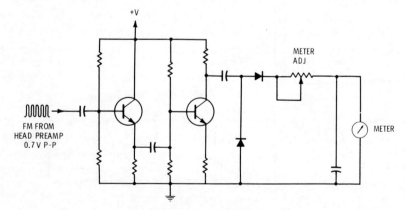

Fig. 10-14. Circuit for tracking meter.

Adjusting the tracking control while viewing the meter will show if the adjustment is correct or not, and this is far better than watching the picture on the screen. Also, a lessening in the playback level will appear on the meter before it will cause a picture breakup on the screen. Not all machines have a tracking meter, and in these cases viewing the monitor must suffice.

11

Editing

The editing of tape means the arranging or rearranging of previously recorded material from the same or different tapes to make up the desired program on one tape. There are various methods available to achieve this: physical splicing, electronic processing, and electromechanical methods.

With single-track audio tape, editing by physical splicing is a relatively simple matter. The tape is merely cut at the appropriate point and joined to the next desired piece of program with splicing tape. The tools required are a marking pencil, scissors or razor blade, and a roll of splicing tape. The skill is in locating the correct place on the tape to cut, and in making a clean splice. An experienced tape editor can accurately remove and replace music passages so that the editing will be unnoticed by a trained musician. It is even possible to remove the middle of a word so that the result cannot be detected by another experienced editor.

With video tape, a different situation exists. Cutting and joining a quad tape is theoretically possible, since the tracks are almost at 90° to the tape length, but this method is now rarely used. With helical machines, the cutting is almost impossible to even attempt. To cut in a slant across the tape so that the cut would run exactly parallel to the video tracks would be out of the question. It would still be very difficult to cut at the correct point so as not to disturb the control track. To cut at an angle other than the video-track angle would produce a bar of noise which would run up the screen for each of the fields represented by the sliced tracks. Also, making a perfect joint after such a cut is impossible because the video tracks cannot be made to align.

A video tape can be developed with a solvent containing minute ferrous particles that align themselves along the video tracks and control pulses, and thus make the recorded patterns visible. But even with this help, the cutting and splicing of the tape cannot be performed so that the picture is undisturbed. Apart from the fact that the picture is not perfect, there is another objection; the physical joint itself can cause problems. If the tape is not exactly aligned, then tracking is lost as the joint passes the heads. A bar of noise is seen to run up or down the picture, depending on which way the heads rotate and which way the tape is wrapped. Video can be lost completely for a short time because the video tracks are out of alignment and the heads have to retrack. But perhaps most serious of all is that the sticky splicing tape can touch the heads and cause them to stick to the tape or pick up excessive dirt, both of which will lead to head damage very quickly.

From the above it is obvious that some form of electronic editing is required. This has been developed over a period of time, and now most machines have some form of editing available. Various degrees of editing complexity exist, and not all models have a full editing capability. So it is important to understand just what editing is, how it is achieved, and the type of facility that is available.

The simplest form of editing occurs when separate segments of a program are being shot in a studio and the end result has to be a complete production without any breaks. The first segment is recorded on tape, and then the second segment is recorded after the first. The object is to make the playback look like a simple cut from one camera to another. This technique is required if only one camera is available, or if an artist has made a mistake that has to be removed. In either case, the previous take is played back up to the point where the mistake occurred, and then, without stopping the tape, the record edit button is pressed and the artist picks up his action and continues. In this way, the program is assembled a section at a time. This type of assembly editing is very common and is the easiest to accomplish electronically and in production. In fact, many tv programs are put together in this manner, and many helical machines have assembly edit capabilities.

A more complicated but similar form involves prerecording all the parts of the program on tape and successively copying sections of various tapes onto a master tape that forms the final program. This is similar to the method described previously, but now a second vtr instead of a studio camera forms the program source. This is still a form of assembly editing. Inputs from separate sources (cameras and vtr's) can be mixed together in an assembly session, and provided some preplanning and thought have gone into what is required, much can be accomplished in this way.

Another situation occurs when a finished program is discovered to have a mistake of some form in the middle. The difference in this situation is that now there is material that must be preserved already on the tape after the new section. In other words, instead of adding a section onto the end of an incomplete program, we now wish to insert a section into a complete program. This is an entirely different situation. Insert editing is not available on most machines. It is limited to the more expensive models because the electronic and mechanical requirements are much more stringent. Also, it is more difficult operationally.

Electronic editing relies on two facts: (1) a tape copy is as good as the original, and (2) it is easy to switch electronically from one signal to another. A combination of these two facts makes it possible to copy the required sections of tape successively in the order required to produce a program as good as the original, and with no obvious breaks. When the first section has been copied, the tape-machine input is switched to the next section to be copied, and so on, until the complete program is on one tape. Obviously, in order to do this, all the video signals must be in step with respect to their sync pulses. It is achieving this that is the essence of video editing.

Electronic editing is usually classified in two categories, *assembly* and *insert*. The editing mode known as *audio dub* will not be covered at this time, since this is really an audio record mode that accompanies the playback of the video. Strictly speaking, it is not an electronic video editing procedure, since the video is left untouched.

On machines with two audio tracks, the operational procedures should be carefully checked because provision is often made for leaving one track untouched while the other is erased or recorded.

The time of switching from the playback to the record mode in video editing also defines the type of editing facility provided. Two situations exist, vertical-interval and random.

There are three methods of achieving the switching to remove the old, unwanted video and to record the new material: (1) main erase of whole tape, (2) increased record-head current, and (3) flying erase head.

Of the above three possible methods of classifying the editing mode or procedure, any one can be found with any of the others. The following examples should make this clear:

Sony AV 3650: Assembly edit, increased head current, random switching.

Sony EV 320: Insert edit, increased head current, random switching.

IVC Machines: Insert edit, flying erase head, vertical-interval switching.

No matter which edit facilities are provided on a machine, there are two functions of importance common to all modes: (1) the capstan must be servo controlled instead of running at a constant speed, and (2) some form of selective erasing is required. The erasing procedure or method is the main difference that distinguishes the assembly and insert modes.

SERVOS FOR EDITING

In order for editing to take place, the incoming video must be in the same time frame as that already on the tape. When this is so, the transition from the playback of the video on the tape to the recording of the incoming new video can be effected without any problems. For this purpose, it is sufficient that the vertical intervals on the tape and the incoming picture be aligned, as shown in Fig. 11-1.

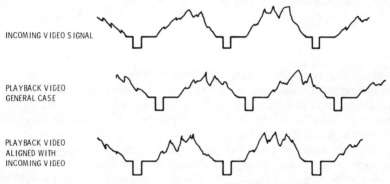

INCOMING VIDEO SIGNAL

PLAYBACK VIDEO
GENERAL CASE

PLAYBACK VIDEO
ALIGNED WITH
INCOMING VIDEO

Fig. 11-1. Alignment of video signals.

To have a prerecorded tape play back so that the sync pulses align with those of an incoming signal means that some control over the tape motion is necessary, since this is the only way the vertical interval of the playback can be moved so that it will be in step with the incoming new video. Because the recorded signal must assume an exact physical position with respect to the rotating head for correct playback, and the head must also be in step with the incoming video, it becomes necessary to control the speed of the capstan as well as the head to achieve this three-way alignment. A machine that does not have a capstan servo cannot be used for editing, because the capstan speed cannot be made to vary and thus align the tape with the new picture being fed into the machine.

In order to obtain the required alignment, the stripped sync of the new video is compared to the control pulses that are already on the tape, and the capstan servo adjusts the speed of the tape until these are coincident. When this has occurred, the head servo adjusts the rotating-head position for correct playback tracking, and then the playback and the incoming video are in perfect sync. When this state has been achieved, it is possible to transfer from the playback mode to recording of the incoming signal. The tape is in the correct position physically and has the correct relationship with the head drum, the rotating head, and the incoming video. This coincidence is illustrated in Fig. 11-2.

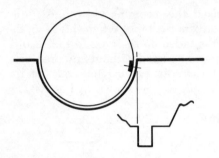

Fig. 11-2. Head and signal coincidence.

Once the changeover from the playback of the old video to the recording of the new has occurred, the record process continues as in normal recording. On playback, all of the control pulses will align perfectly, and the tracks of the new video will lie alongside those of the old with no gaps, bunching, or incomplete pictures.

It is important to realize the difference in the servo modes in the two forms of editing. In add-on or assembly editing, the function is to enter the record mode at some given point and then remain in that mode. Thus the servos work in the playback mode until the edit button is pushed, and then they change over to the record mode. In insert editing this is not so. The servos begin in the playback mode and remain in this mode. At no time do they enter the record mode of operation. The servos continue to lock up on the incoming video and the control tracks from the tape, which are not erased. The only parts of the machine that enter into the record mode are the appropriate erase sections and the record head, both of which revert to the playback mode when the end-insert or cut-out button is depressed.

EDITING ERASE

The erase functions in editing are of vital importance and of varying complexity; they are something that must be controlled perfectly. In all machines, the main erase head is positioned ahead of the drum assembly so that the tape is erased before it comes into contact with

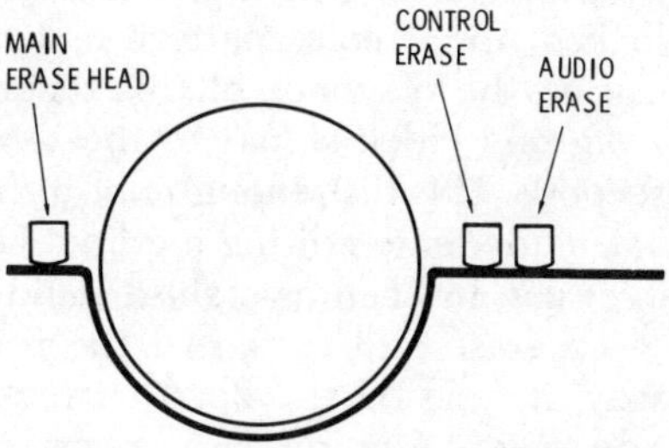

MAIN
ERASE HEAD

CONTROL
ERASE

AUDIO
ERASE

Fig. 11-3. Locations of erase heads.

the rotating head. This erase head is usually wide enough to cover the video tracks and the control tracks; the audio is erased by its own head. Fig. 11-3 shows the approximate positions of these heads.

Main Erase of Entire Tape

In Fig. 11-4, it can be seen that the tracks from B onward have an increasing portion of their length erased. If these tracks are played back, the loss of video will crawl up or down the screen with each successive frame. If this loss of video were replaced by continuing the video track with new information, selection of the varying time of changeover from track to track to produce a gradual creeping in of the new picture would be both impossible and undesirable. Therefore, the main erase head is not used in this fashion for editing. It is used to clean the tape of the old video, but in conjunction with the record circuits and other erasing methods in such a manner that perfect editing can result.

Increased Head Current

An easier and better way to supply new video to the heads is to start at track A in Fig. 11-4. When this point is reached, if the record current to the heads is increased to about 125 percent of that used for normal recording, it will effectively obliterate the previous video and impress a new signal. The time from point A to point Z is about 3 to 3½ seconds in most machines, and the new record current is set to last this length of time before it drops to the normal level (beyond point Z, the tape is completely erased by the main head). This

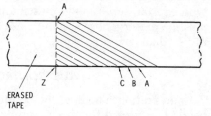

Fig. 11-4. Erasure of tracks.

A

Z   C B A

ERASED
TAPE

process can be accomplished quite easily, but if the servo is not locked up, a moire pattern or parts of the old picture will be seen during the playback of this section of the tape. The switching time from one video signal to the next is accomplished in a few microseconds. It is fast enough not to be visibly discerned and thus serves as an adequate editing method. On still-frame playback, the changeover can be seen as a short band of noise in one picture line.

To erase and re-record the tracks on an individual basis in this way, increasing the head current will work, but it has a major disadvantage. The increase in current almost saturates the tape, so that if a mistake occurs, a second chance to correct it will not be possible. The record current will now not be strong enough to override the video, and picture chaos will result. If the second and even third attempts can be made, there will still be a limiting point beyond which this technique will not work. To ensure that repeated attempts will be successful, another method is adopted.

Flying Erase Head

A head similar to the record head can be mounted a few degrees in front of the record head on the rotating plate, and an erase current can be fed to it. Such a head is known as a *flying erase head.* This new head scans the video tracks just before the main record head, so if it is fed an erase current it will affect only one track at a time. The erase current is timed to start just before the erase head begins a scan, and the new video is applied to the record head at the same time. Removal of the erase current is concurrent with the removal of the new video so that only the selected tracks are re-recorded. Now only the exact portions of video that are required to be replaced are erased and re-recorded. This allows editing mistakes to be corrected easily as often as necessary. This method is most suitable for insert editing, and it is found only on machines with such capability.

When a flying erase head is used, it must track the recorded tracks exactly. A slight wandering of the tape speed or heads will cause the flying erase head to mistrack and thus not erase fully, and the new tracks will not be exactly over the paths of the old tracks. This will obviously cause problems on playback. For this reason, a playback-type capstan servo mode must be used, and thus the control tracks must be preserved, so a flying erase head is suitable for insert editing rather than assembly work.

THE CONTROL TRACK

In an editing machine, the control track must have its own erase head. The reason for this is simple and can be visualized with the

218

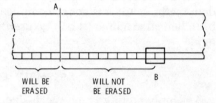

Fig. 11-5. Control-track erasure.

WILL BE ERASED WILL NOT BE ERASED

help of Fig. 11-5. The main erase head begins to erase the tape fully at point A. The control-track record head is at point B. Between these two points is a length of tape with control-track pulses recorded on it. The control head starts to lay down new control pulses when the edit button is pressed, so for this length of tape there will be a double set of control pulses which may not be coincident. On playback, these would certainly upset the servos for this section of tape. This problem can be avoided by erasing the control track just before the control record head, leaving only one set of pulses along the entire length of the tape.

A machine with assembly editing only can have a main erase head which will cover the control track, but in the insert mode the control track is not erased and re-recorded. In this type of machine, the main erase head will not extend over the section of the tape reserved for the control track. Hence, a separate control erase head is required on machines with insert capability.

RECORD AND ERASE CURRENT SWITCHING

There are two methods of switching from the old to the new video. This switching can be done at random, or it can be restricted to the vertical interval.

Random switching means that the switch to the new video and the application of the erase current occur when the button is pressed, which can occur at any time during the picture. Provided the actual switching is effected electronically, which is generally the case, this method provides an acceptable edit. The changeover time is typically a few microseconds and cannot be pinpointed by the eye on playback. It is an easy method to implement without complicated electronics or excessive cost, and many helical machines use it with considerable success.

Vertical-interval switching means that the switchover takes place during the vertical interval that occurs after the operate button is fully depressed. This method has the advantage that it cannot be seen on the screen, no glitches or other interference will appear, and any other disturbances are less likely to affect the video signal. This is, of course, the accepted broadcast method.

Vertical-interval switching requires more electronics than the random method, but it is not so complicated as to be out of the question for a helical machine. It can be used in both the assemble and insert modes.

ASSEMBLY EDITING

Assembly editing is the successive adding of video sections on the end of the previously recorded video. After the cut to new video is made, the previous material is not returned to, except in playback. Assembly is the easiest of the two modes and is often provided on machines that do not have the insert mode.

When two sections have been recorded on tape, the playback should look like a smooth cut from one camera to another and should be free from rolling, jumps, or picture tearing. Obviously, as the new video is recorded, anything that previously was on the tape must be erased. Since the assembly process is just like normal recording once the button has been pressed, the main erase head is operated, and the whole tape is cleared before it enters the scanning drum. Hence, the old video, audio, and control-track recordings are eradicated. On stopping the machine and playing back an assembly edit, the end point of the edit is seen as an increasing band of noise crawling up the screen.

During the three seconds or so that it takes the tape to travel from the main erase head to the scanning head, there is a section of tape that will be recorded over but which has not been erased by the main erase head. Therefore, either the extra-head-current method or a flying erase head must be used for this length of time.

In this fashion, it is possible to add new program segments of any length, from as short as a few seconds to almost complete program lengths. Since a new control track is recorded in exact step with that already on the tape, the finished product looks like a continuous complete recording, which is exactly what is required.

The new series of control pulses is recorded at the same time that the new video is laid down, the switch being made at the same time. These pulses begin just before point A in Fig. 11-6, and on playback they line up with the previous pulses. In this way, if the video or

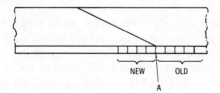

Fig. 11-6. Old and new control pulses.

NEW   OLD

A

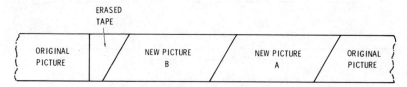

Fig. 11-7. Example of assemble edits.

head or tape drifts slightly, the servo correction can take effect, and this will be reflected in the pulses. If the previous track was preserved, the playback video would not necessarily align perfectly with the pulses, and mistracking could occur.

Fig. 11-7 shows the layout on the tape of two successive assemble edits that have been added to a base picture.

INSERT EDITING

Insert editing differs from assembly work in some important respects. The main difference is that at the end of the insert of the new video it is necessary to cut back to the old material already on the tape. This makes necessary several changes in the electronic and mechanical provisions of the machine.

In assembly work, the minor variations that may occur between the new incoming video and that already on the tape can be tolerated after the edit record mode has been entered, since there is no return to the old video on the tape. In this mode, a new control track that corresponds to the new video is put onto the tape.

In the insert mode, these minor variations cannot be accepted because a return to the material on the tape is to be made, and perfect tracking must be maintained at both the cut-in and cut-out points. For this reason, the servo maintains a very tight lock, and a new control track is not recorded—the existing series of control pulses is retained. It is this more critical circuitry with regard to the erase and servo functions that distinguishes the insert machines from the others.

Because a new control track is not recorded and a return to the original video is made, an entirely new erase system is required. A main erase head that scans the whole width of the tape obviously cannot be used. The tracks must now be erased and recorded on a track-by-track basis. If this were not done and the main head were used, a band of erased tape would exist between the last tracks to be inserted and those tracks that had not yet reached the full erase head.

Insert editing can be effected with either the increased-head-current method or a flying erase head. In both cases, the full erase

Fig. 11-8. Example of insert edits.

head cannot be used. Random or vertical-interval switching can be used with equal success. Fig. 11-8 shows how the inserted sections fit together on the tape.

Usually it is required to keep the audio in this mode; hence the audio track has its own erase head. Often, audio is added at a later time, after the video has been finalized, so the audio-dub mode is a necessity with insert editing. It is common to find insert edit machines with two audio tracks. These should be carefully checked, because one track is erased and re-recorded every time new video is put onto the tape, while the other is left untouched.

THE STOP BUTTON

If a mistake occurs during an edit, the process can be stopped by using the stop button. In the assemble mode, this is the normal method of ending the new section, but it leaves a blank between the instantly ended new video and the old video on the tape. This section of tape is unsuitable for playback and must be recorded over in the correcting of the mistake or in the adding of the next section. This blank is due to the section of tape that has passed the main erase head but has not yet reached the scanning heads (Fig. 11-9). It has no video or control track on it.

In the insert mode, the situation is less severe. At best the stopping point can look like an ordinary cut-out, and at worst there will be about one third of a field erased by a scanning erase head. Also, depending on how long it takes the tape to stop, there could be a small section of tape with a few control pulses missing. In this mode, the safest procedure is to use the cut-out button and then press the stop button a few seconds later.

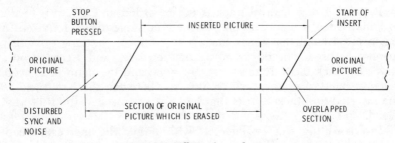

Fig. 11-9. Effect of stop button.

AUDIO

In both of the edit modes, the audio must be carefully considered. The audio head is on the opposite side of the drum from the main erase head, and the first new video track is placed about midway between the erase and audio heads (Fig. 11-10). The audio erase head, if a separate one is used, is usually mounted in the same housing as the audio record head, or just in front of it in another, similar mounting.

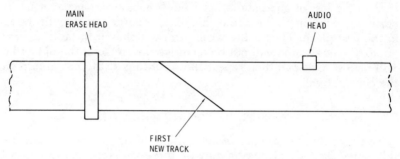

MAIN
ERASE HEAD

AUDIO
HEAD

FIRST
NEW TRACK

Fig. 11-10. Erase-head and track positions.

Due to this physical difference in placement, there is a small but important time difference between the audio and video erasure points. Thus there is no alignment between the erasing and recording of new video and the audio which corresponds to these particular frames. The audio is physically offset along the length of the tape exactly as in a film. This means that the beginning of an edit on a visual cue can cause a problem with untouched audio, since the time difference between the main erase and audio heads is about 3 seconds. In the most simple of the assembly modes, the audio situation is just as described, and in production this must be considered.

To compensate for this situation, most machines have an audio-dub facility. This enables the audio to be recorded separately after the video has been completely assembled. An audio-dub button is provided which powers the audio record amplifier and provides bias to the audio erase head only. This head is mounted just ahead of the audio record/reproduce head and spans the audio track only. The vtr is placed in the normal playback mode, and when the audio-dub button is pressed, video playback continues, but recording takes place on the audio tracks.

The audio board or section of board in most vtr's is powered or switched separately from the video or servo boards. Hence, putting the audio section into the record mode with the rest of the machine in the playback mode is very easy. The audio-dub push button simply operates the audio-board slide switch or powers only the relay or solenoid for the audio. In effect, the recorder becomes a simple audio recorder for adding an audio accompanyment to a picture viewed on a screen.

When insert editing is being done, care should be taken to see what happens to the audio. Most machines with this capability have two audio tracks, one of which is left untouched while the other is activated for new audio. This arrangement has definite production advantages, but confusion as to which track is which can lead to the loss of good audio. As a safety measure, whenever insert editing is envisoned, the audio should be recorded onto both tracks in the first place. Thus if the wrong track is chosen for adding new audio, the new audio will not be recorded, but none of the old audio will be lost. One can then repeat the recording of the new audio on the correct track.

EXAMPLES OF EDITING FACILITIES

To illustrate the methods of achieving editing in a helical vtr, the following sections give examples of three different machines. All have different but representative ways of accomplishing the desired end.

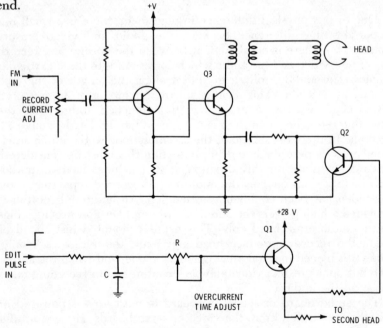

Fig. 11-11. Simplified Sony AV 3650 edit circuit.

Sony AV 3650

The Sony AV 3650 is equipped with the assemble mode only. It uses the increased-head-current method and random switching.

The edit-in pulse from the edit button turns on Q1 (Fig. 11-11), which is then held on by the charge on capacitor C for a time determined by potentiometer R. This turns on Q2, which shorts out the emitter resistor of Q3, thus increasing the gain of Q3 and supplying more current to the record head. The head current is set to remain increased for about 3 seconds from the moment the button is pressed.

The servo in the playback mode is locked to the vertical sync of the incoming video so that both video signals are in phase. When the edit button is pressed, the drum servo remains unchanged, with the head-tach pulse compared to the vertical-drive signal in both playback and record. The capstan phase servo uses the control pulses from the tape and the head tach in playback; it switches to vertical drive and head tach in record. This change is made in a fraction of a second, and no loss of servo lock-up occurs.

This format has only one audio track, which is re-recorded with the video when the edit button is pressed. An audio-dub facility is provided and is operated by a separate button.

Sony EV 320F

The Sony EV 320F is provided with both assembly and insert editing. It uses a flying erase head and has both random switching and simplified vertical-interval switching. Two audio tracks are provided, one of which has a separate audio-dub capability. (It should be noted that facilities vary from model to model.)

The early models had a flying erase head that was fed with a boosted modulated rf signal to erase the tape. This rf signal was the normal record signal fed to the record head, but was amplified to the point where it served as a suitable erase current. It was used in both the assemble and insert modes, and in neither case was the record-head current increased. In the assemble mode, both the main erase head and the flying erase head were turned on, but the flying head was turned off after about three seconds when erased tape began passing around the drum. In the insert mode, only the flying head was used. In both modes, random switching was used.

The later models are somewhat different. Basically, the setup is the same as in the earlier models, but there are two main differences. The flying head is now fed with an unmodulated rf signal from its own separate oscillator, and a modified form of vertical-interval switching is used.

Instead of using the incoming vertical-sync pulse, the vertical-interval switching is effected indirectly. A new coil is placed in the

225

head drum assembly to sense the position of the rotating head and to create an edit pulse at the correct time. Fig. 11-12 shows the position of this coil in relation to the normal head-tach coil, and also shows the relative positions of the flying erase head and the record head.

When the rotating erase head first contacts the tape, it scans the

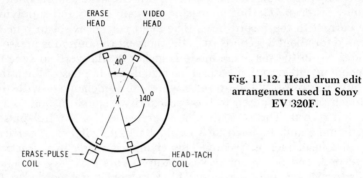

Fig. 11-12. Head drum edit arrangement used in Sony EV 320F.

vertical-blanking interval. The edit coil generates a pulse at this time, and the pulse initiates the flow of erase current to the head. Video current is applied 3.2 milliseconds later when the record head contacts the tape in the same vertical-blanking interval. Note that this vertical-blanking interval is the same for both the video on the tape and the incoming video. The new recording is thus started during this interval.

Fig. 11-13 is a simplified block diagram of the process. The edit or cut-in button sets up one side of gate circuit A when it is depressed. This allows the next pulse from the edit coil to pass through and change the state of flip-flop B. This turns on gate C and allows the rf current to reach the flying erase head. The flip-flop also triggers monostable D, which 3.2 milliseconds later changes flip-flop E and thus allows record current to reach the record head through gate F.

At the end of the edit sequence, the end or cut-out button is pressed. This closes gate A and resets flip-flop B. Immediately, the erase current is stopped, and 3.6 milliseconds later monostable G resets flip-flop E and cuts off the video record current.

The cut-in point should occur between 2 and 4 horizontal lines after the end of the vertical-sync pulse, and an adjustment is provided in the machine to set this up. Fig. 11-14 shows this place in the video waveform.

This has been an extremely simplified description of this editing circuit and has left out most of the circuit details and some of the available facilities. However, the service manuals cover this infor-

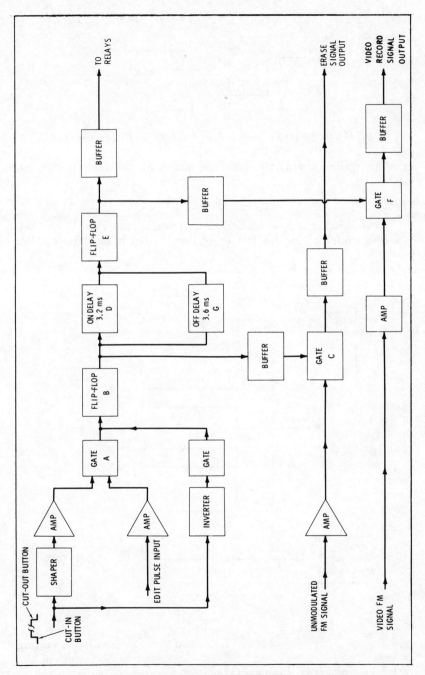

Fig. 11-13. Block diagram of Sony EV 320F vertical-interval edit circuit.

227

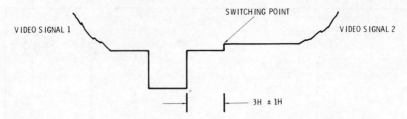

Fig. 11-14. Switching point in Sony EV 320F vertical-interval edit.

mation quite adequately and are sufficient for normal use and servicing.

IVC

The following is a simplified description of the IVC Model 960. In this model, there are five erase heads, which must be set into

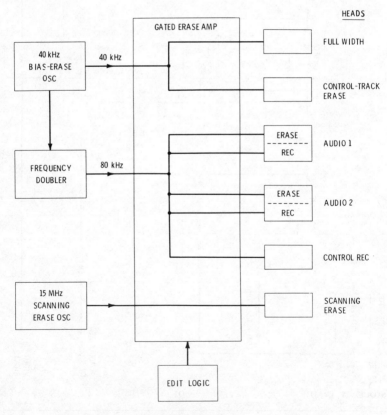

Fig. 11-15. Simplified block diagram of IVC edit system.

228

operation at the correct times to perform the required erase functions. These are shown in Fig. 11-15 along with the signal that feeds each head. The scanning erase head is fed a 15-MHz signal from a special oscillator used for this purpose only, while the other heads are fed from a 40-kHz oscillator through a gated erase amplifier. This amplifier is gated on by a signal from the editing logic board, which is ultimately controlled by the operation of the front-panel push buttons.

The next paragraphs describe a few of the facilities of this machine. This description will be brief and will not attempt to explain the circuit operation. The service manual is complete and comprehensive and cannot be condensed without sacrificing essential material. No vtr should ever be serviced or modified without full reference to the appropriate manual for that model.

Assemble Mode—The assemble mode involves a vertical-interval switching operation in which all the erase functions are identical to those in the normal record mode, except for the addition of the scanning erase head. The rf erase signal is fed to this head through a rotating capacitor mounted in the head-drum mechanism. The erase rf is activated for about 3 seconds.

Insert Mode—Of the video erase functions, only the scanning erase is used in the insert mode. It is activated 250 milliseconds after the mode is initiated and remains active for the duration of the insert. When the insert is ended, the erase is timed to continue for another 650 milliseconds, and then the machine reverts to its normal playback mode. The control of these timed periods is one purpose of the edit logic board.

Of the two audio tracks, track 1 is inhibited, and track 2 is selected if required by a front-panel push button. Various modes of audio recording and editing are selectable, and the operation manual should be consulted for these.

The Scanning-Erase Oscillator—The scanning-erase oscillator is contained in one IC and produces a 15-MHz signal. The output is transformer coupled to a rotating capacitor in the head assembly, and it is used to feed the scanning erase head only. It is used on editing modes only and not in the normal record operation.

Normal Record Mode—When the normal record mode is initiated, the full-width erase head and the control-track erase head are immediately fed with a 40-kHz signal from the erase oscillator. The output of this oscillator is also doubled to 80 kHz. This signal is fed immediately to the control-track record amplifier as a bias signal. It is used because it allows a minimum of adjustments when tape stock is changed. After a delay of 250 milliseconds, the 80-kHz output is passed by the gated erase amplifier and appears as bias and erase signals for both audio tracks.

Edit Logic—The edit logic is an extremely complicated logic and timing circuit, designed to initiate the desired modes in the first vertical interval after a delay period started by the push buttons. Everything is controlled by a counter circuit and a series of coincidence gates, all of which are contained in various ICs. Several delays are used to allow the transport to settle and so that the scanning heads can reach the correct speed and position for editing. As mentioned previously, the manual contains excellent descriptions and circuits for servicing, and a shortened version here would serve no useful purpose, particularly if the reader does not have access to a machine to study.

Auto Framing—Auto framing is a feature provided only on the top line and the most expensive of the helical vtr's. Its purpose is to ensure that the incoming video and the playback signal are in the same field and not separated by a half line at the time of an edit. It examines the two signals and ensures a frame match before an edit can occur.

Field 1 starts with a full line and terminates with a half line, whereas the opposite is true for field 2. Thus the vertical sync at the end of each field is displaced by a half line. This half-line displacement is used in a logic circuit to identify the individual fields and to provide the auto framing. In operation, it is similar to the lock detector in the capstan servo. This detector indicates a frame match on a relatively few frame coincidences, but it indicates a loss of match after several coincidences have been missed. It is desirable to secure frame match during edit play as well as during normal play but to inhibit this track shifting or framing action during the record mode.

A special problem exists at the end of an insert when the machine reverts to the normal playback mode without stopping the tape, since an unmatched condition could occur at this time. To prevent a track shift at this point, a delay that causes a perfect action of the machine is used.

Circuits are also included to prevent frame hunting or loss of match at the head-dropout time, and to prevent triggering by low video and noise at the demodulator output if the pim falls to a low level.

EDITING ACCESSORIES

Several accessory or add-on units are available for the purpose of making editing quicker and easier to accomplish. These devices fall into two basic categories, those that require a special time code and those that make use of the control-track pulses. These are described in the following paragraphs.

Editing With SMPTE Code

In one type of editor, the add-on modules are used to record a series of coded pulses onto the cue or second audio track of the tape. This pulse code, known as the SMPTE* code (Fig. 11-16), uniquely identifies each frame by the frame number, a time in hours, minutes, and seconds, or some other arbitrarily chosen information. The code uses 80 binary bits of information in each frame. Sixteen of these bits form a sync word to indicate direction of tape travel and end of frame. The code requires no external timing circuits in its playback process.

The code has the capability of handling frame counts in both monochrome and color systems. In monochrome, 30 frames per second are used, but in NTSC color only 29.97002618 frames per second are used. This slight difference is enough to require two frames to be dropped every minute, except in the tenth minute. All of this is taken care of automatically by the circuits in the modules.

Recording the code can be done at any time before, during, or after the original recording. Once the code is on the tape, every frame is uniquely identified by the information in the code, and there is no need to count sprocket holes, feet, or other pulses.

The following modules are available:

Time Code Generator
Auxiliary Display
Edit Code Generator
Electronic Editor Programmer
Control Panel
Auxiliary Programmer
Video Character Generator
Edit Code Reader
Synchronizer

All of these allow full automation of all the edit functions; this includes generating and recording the code, cuing, edit-point selection, previewing, and playback. Each frame can be visually identified by a readout, and the required frame can be manually set on another readout. Any chosen sequence or single frame can be found and viewed rapidly with single-frame accuracy, and the modules can even take two vtr's that are 30 seconds apart in program time and bring them into synchronization for editing. All of the edit functions can be preset to within one frame, and the modules will then perform the cut-in and cut-out or changeover from one machine

---

*SMPTE is the abbreviation for Society of Motion Picture and Television Engineers.

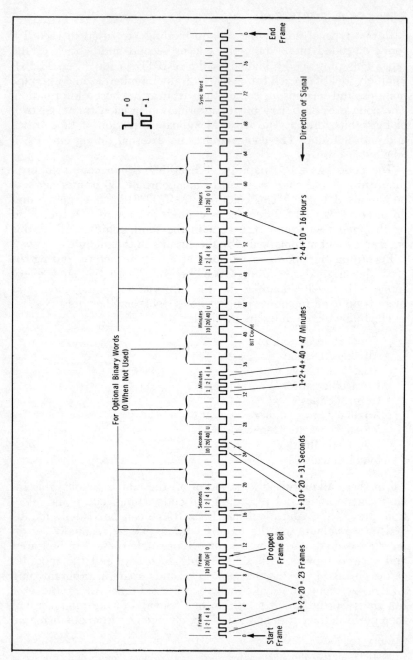

Fig. 11-16. SMPTE edit code.

Courtesy Electronic Engineering Co. of California
Fig. 11-17. EECO editing modules.

to the other. The modules contain complex electronic circuits that monitor the code and behave in much the same manner as a small specialized computer.

This equipment was designed mainly for broadcast use, where considerable time and equipment were tied up in editing major network shows. If this time could be reduced, so could the substantial costs of editing. Fig. 11-17 shows a system, manufactured by Electronic Engineering Company of California (EECO), for use in studios engaged in network production. A selection of simpler equipment is available for use with helical machines in nonbroadcast applications.

Editing With Control Pulses

Another type of editor, for use with smaller helical machines and U-matic cassette machines, uses the control pulses instead of the SMPTE time code. A typical setup is to have two vtr's connected to the editor by multiconductor cables; one machine is designated the recorder and the other the player. The cables carry the control pulses to counters in the editor, and they also make connections that allow the major operational functions to be remotely controlled at the editor panel.

The editor unit includes circuits to perform functions not normally available on the machines themselves. For example, they permit low- and high-speed playback in both directions, and simultaneous preroll of both machines to set up an edit.

In operation, the beginning of the tape is found, and the counter-timers for both machines are set to zero. Then the tapes are individually searched to find the desired edit points. When these points are reached the machines are stopped, and the preroll button

233

is used. This will automatically rewind both tapes to a point 5 seconds before the edit point. Then both machines are put into the play mode. At the edit point, the record machine is switched to the record mode automatically, and the edit is made. An edit can be previewed before it is made. To do this, both machines are rolled, and the edit is displayed on the record-machine monitor, but the machine is not actually put into the record mode. If the edit appears satisfactory on the monitor, the preroll can be repeated and the edit actually made. If the edit needs adjustment, the counter-timer for either machine can be advanced or retarded to make the edit point correct.

The more advanced editors allow the edit in and out points indicated on the timers to be entered into a memory, and a programmed series of edits can be made automatically.

This type of editor relies on reading the control pulses in playback, fast forward, and rewind. Therefore, a stable and positive tape drive is needed in all modes.

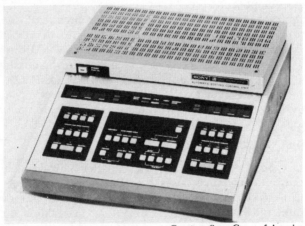

Courtesy Sony Corp. of America
Fig. 11-18. Sony BVE-500 editing control unit.

Some of these editors have been manufactured to work with one particular model of vtr only; an example is the BVE-500 (Fig. 11-18), which is used with the U-matic machines. Others can be used with more than one type of vtr. In these, the main circuits of the editor remain unaltered, and an "interface" board is used to permit a variety of vtr's to be used.

The edits made with these editor machines are accurate to within a frame, and these editors are used regularly in documentaries and network news programs.

MECHANICAL SPLICING

Many people experienced in the field of video tape do not recommend the physical cutting and splicing of the tape. However, there have been times when this has been unavoidable, and to cover these the advice given by Ampex is best heeded. It is reprinted here by permission:

1. The edges of the two pieces of tape must be precisely aligned so that there is no lateral displacement at the splice and so that the two pieces of tape are parallel to each other. Failure to comply with this requirement will, at a minimum, increase the duration of the picture disturbance and may cause the recorder to stall.
2. The ends of the two pieces of tape must be butted together with a maximum of 0.10 inch clearance and without any overlap.
3. A proper splicing tape must be used. The splicing tape is applied to the back, or nonoxide, side of the tape only. This is the side of the tape that does not contact the recording heads.
4. The splicing tape must be trimmed so that it does not extend beyond the sides of the tape. It is permissible to undercut the edge of the tape at the splice area to meet this requirement.

There are several tape splicers on the market that will meet requirements 1 and 2. These are available at radio/tv distributors. Splicing tape should be preferably ½ inch wide, 0.0005 inch thick, and aluminized.

12

Color Recording and Playback

Two methods have been used successfully to record and play back color with video tape. The first is the "direct" method used in quad and broadcast helical machines, and the second is the down-converted subcarrier, or "color under," method used in the smaller helical machines. To put color recording and playback into perspective, it is easier to treat the direct method first.

THE DIRECT METHOD

If the full NTSC color signal is applied to the input of a vtr, it will be recorded onto the tape just as a monochrome signal will. The color signal can be demodulated and reproduced just as a monochrome signal can, and there might seem to be no reason why this method should not be used. However, there can be serious problems with sidebands when the signal is demodulated.

An example can be used to show how these problems arise. Consider a 5.9-MHz carrier frequency modulated with a standard color signal. The sidebands caused by the color subcarrier can be calculated as follows:

1st order sidebands: $5.9 \pm 3.58 = 9.48$ or 2.32 MHz
2nd order sidebands: $5.9 \pm 2(3.58) = 13.06$ or -1.26 MHz
3rd order sidebands: $5.9 \pm 3(3.58) = 16.64$ or -4.84 MHz

The two "negative" frequencies "fold back" and do in fact appear as real positive frequencies. These interact with the carrier to produce frequencies of $5.9 - 1.26 = 4.64$ MHz and $5.9 - 4.84 = 1.06$ MHz. The 4.64-MHz frequency is easily filtered out of the video signal, but the 1.06-MHz frequency will cause objectionable beats and moire patterns in the picture.

There are two satisfactory ways to avoid this problem. The first is to choose a carrier frequency that will not produce the problem. This is easy to do, but the necessary frequency is rather high. For example, the standard broadcast "high-band" fm range is 7.06 MHz to 10.0 MHz. To record such frequencies satisfactorily, a writing speed of at least 1000 in/s is required. Quad and broadcast helical recorders have a high enough writing speed, and this method is adopted for color recording with these machines.

The second method of avoiding the problems is to use the "color under" method. This method must be used in the small helical machines that do not have a high enough writing speed to use high-band fm.

COLOR PLAYBACK PROBLEMS

It is important to realize that *no* vtr will play back a color signal perfectly. All exhibit unstable color on a screen, and instability can be seen in the signal when it is viewed on an oscilloscope.

Several factors contribute to incorrect playback of the color. Perfect playback would occur if the tape were subjected to exactly the same conditions on playback as it was on record, but this is not usually the case. The main problems that affect the color are:

1. Tape stretch
2. Wow and flutter in the tape transport
3. Tension changes in the tape
4. Slight servo instabilities and normal servo corrections
5. Noise problems in the signal
6. Beats in the picture

The effects of these parameters can easily be cancelled out in monochrome, but in color they alter the phase or frequency of the recorded 3.58-MHz signal. Tape stretch, for instance, will lower the frequency slightly, and speed instabilities will vary the frequency up and down. The result is changes in hue or complete loss of color, or irregular color bands across the screen.

To illustrate how easy it is for color problems to occur in a helical vtr, consider first one cycle of a 3.58-MHz signal. Fig. 12-1 shows how the successive peaks of the signal occur at points separated in time by 0.279 microsecond. For perfect color playback, any changes in this signal must be kept to a minimum. A change of 5° will pro-

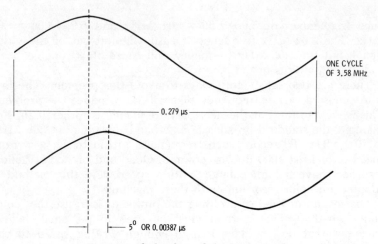

ONE CYCLE
OF 3.58 MHz

0.279 μs

5° OR 0.00387 μs

Fig. 12-1. Slight phase shift of two signals.

duce a noticeable visible effect on the screen. A change of 5° is a timing error of 0.00387 microsecond.

Now consider a helical machine with one head. This head describes a full revolution of 360° in 1/60 second. Thus, in one microsecond it rotates through about 1.3 minutes of arc, and in 0.279 μs it moves through 0.362 minute of arc. So a 5° change of signal is equivalent to 0.005 minute of arc in the rotation of the head.

With a large rotating head mass as in a helical machine, it is unlikely that instabilities of this order would occur. When they do happen, they are of a much larger order, and hence can thoroughly upset the color. Although the servo correction in a helical machine is excellent, it is not this good. A typical speed regulation of 0.002 percent represents corrections in the order of 0.4 minute of arc. So a minor phase variation in the color signal is well beyond the limits of the servo. Hence, some other method of correction for the color problems is required.

The effect of these problems, usually referred to as *time-base errors,* is to alter the time between the successive signal peaks on the tape, i.e., to alter the phase and frequency relationships. It is thus necessary to add some correction circuitry to the simple recording and playback circuits.

The worst of the color errors are caused by tape stretch along its length. In quad machines, the tracks run at almost 90° to the tape length, and so the quad format is not seriously affected by tape stretch. The helical tracks run almost parallel to the tape length, however, and are thus greatly affected.

Wearing down of the heads and slight instabilities of the head-wheel also affect color playback, and in quad machines they make electronic correction of the signal necessary. In helical machines, the same effects appear, but they are not as serious a problem as tape stretch.

The playback instabilities are usually expressed as the amount of timing variation or *jitter* in either the horizontal sync or subcarrier. The servos in a quad or broadcast helical machine will reduce this to about 1 microsecond, which is good enough for monochrome work but is far too great for color. A small helical machine will have several horizontal lines of instability, *i.e.*, 3 or 4 times 63.5 microseconds, which is far too great even for monochome playback.

In the quad machines that use direct color, two stages of color correction are applied. First the signal is sent through an electronically variable delay line to reduce the timing errors to about 30 nanoseconds; then a second line reduces the errors to about 10 nanoseconds.

In the small helical machines, the sync errors are too large for a variable delay line, so other methods must be used. Time-base correction is covered in Chapter 13. The common methods of color correction in the small helical vtr are described later in this chapter.

DIRECT RECORD COLOR CORRECTION

Recording the full NTSC signal onto the tape is the method used in quad recorders and several high-quality helical machines. The early quad methods of color correction were also tried in some of these helical vtr's, with varying degrees of success. Basically two methods of correction were used:

1. Demod-remod methods
2. Heterodyne methods

In both of these methods, the complete NTSC signal is applied to the fm modulator, and the color is pre-emphasized 6 to 8 dB with the higher frequencies of the luminance part of the signal. On playback, the amplitude distortions are removed by equalization, and the signal is applied to the fm limiter chain exactly as in monochrome playback. The final demodulated signal is the full NTSC signal with all time-base errors from the tape.

At this point, the luminance and chrominance are separated by filters. The difference in the two methods is in how the playback color signal is now treated.

Demod-Remod Methods

Two variations of this method of removing color timing errors have been tried. They are:

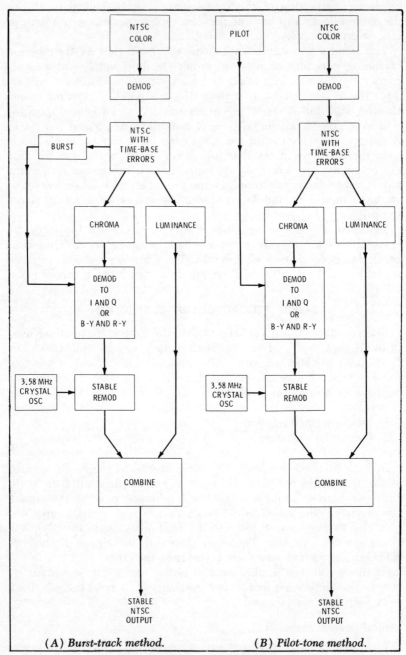

(A) Burst-track method.   (B) Pilot-tone method.

Fig. 12-2. Playback of recorded NTSC signal.

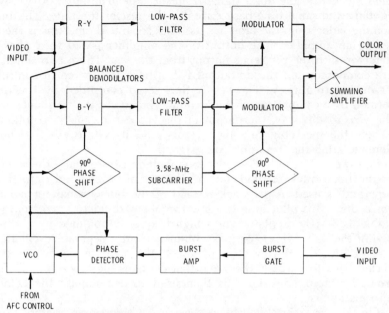

Fig. 12-3. Basic block diagram of burst-track method.

1. The burst-track method
2. The pilot-tone method

Fig. 12-2 shows a block diagram of each method.

Burst-Track Method—In the burst-track method (Fig. 12-3), the color signal is fed to two identical balanced demodulators. The other input to the demodulators is a 3.58-MHz continuous-wave signal containing the same time-base errors as the signal from the tape. This 3.58-MHz signal is phase shifted 90° before it is fed to one of the demodulators so that the output of one demodulator is the R − Y or the I signal, and the output of the other is the B − Y or the Q signal.

It is necessary to preserve the dc components of these signals, so they are passed through low-pass filters. The outputs of the filters are fed to two identical modulator circuits. The other input to these modulators is a stable 3.58-MHz signal, which is fed 90° out of phase to one modulator. This causes remodulation of the color signals, which are now stable enough to be fed to a mixer and then added to the luminance.

The reference signal sent to the demodulators is a continuous-wave 3.58-MHz signal with the same time-base errors as the tape playback. It is derived from a voltage-controlled oscillator that is

241

phase locked to the burst signal from the tape. The burst obviously contains the same time-base errors as the rest of the tape. The incoming video from the tape has its burst separated, and this is then fed to an amplifier. The output of the amplifier is fed to a phase detector with the reference output from the vco. These two signals are compared, and the dc output is fed back to the vco to control its output frequency. The vco is thus servo controlled by the dc formed by a comparison of the incoming burst and the frequency of the vco. In this way, the output of the vco can be made to follow closely the incoming time-base errors, and it is thus suitable for demodulating the incoming color signal.

Pilot-Tone Method—The pilot-tone method is essentially the same as the burst-track method, but a pilot signal is recorded onto the tape and is used in playback instead of the burst signal for color correction. This pilot tone is a low frequency divided down from a stable 3.58-MHz oscillator (or a high frequency obtained by multiplying the 3.58 MHz), or it is obtained from a separate oscillator included for this purpose. On playback, the pilot frequency is converted back to a 3.58-MHz signal that contains the time-base errors from the tape. This signal is then used to demodulate the color information.

Fig. 12-4 is a simplified block diagram of the Ampex method. In the record mode, the burst signal is separated from the video input and then phase compared to a crystal-controlled 3.58-MHz oscillator. The burst and the oscillator output are fed to a phase detector, the output of which controls the frequency of the oscillator through a reactance control circuit. The phase-locked oscillator output is divided by 7 in a blocking-oscillator circuit, the pulse output of which is converted to a sine wave in a tank circuit. This sine wave is recorded onto the tape as a 511-kHz pilot tone.

On playback, the pilot tone from the tape is separated and filtered from the video and amplified until it is a square wave of 511 kHz. Square waves are rich in odd harmonics, so filtering of this signal will produce any odd harmonic required. The filter is chosen to pass the seventh harmonic, which is 3.58 MHz. This output, of course, contains all the time-base errors, and after amplification it is used to demodulate the signal from the tape into the two color signals. These are now remodulated with a stable 3.58-MHz oscillator to give a good NTSC signal.

The Heterodyne Method

The heterodyne method was used in some early quad machines, and also in the earlier Panasonic helical vtr's, such as the NV 3020C. In common with the previous methods, the complete NTSC signal was recorded onto the tape by the fm modulator. On playback, the

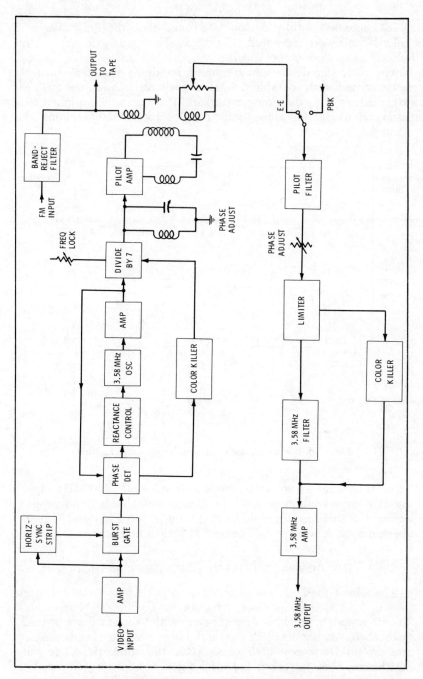

Fig. 12-4. Block diagram of Ampex pilot signal system.

243

NTSC signal was fully demodulated, and then the luminance and color signals were separated (Fig. 12-5). The luminance was treated exactly as in monochrome work.

The color signal was sent through a bandpass amplifier and then heterodyned with a stable 5.4-MHz oscillator to produce an 8.98-MHz carrier with the time-base errors. This signal was applied to a balanced modulator along with a 5.4-MHz signal containing the

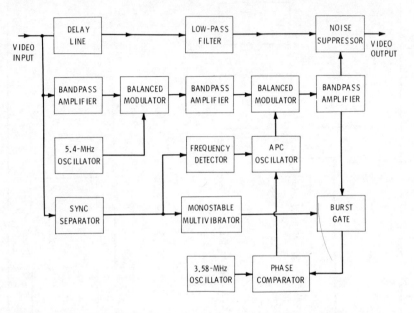

Fig. 12-5. Block diagram of early color playback method.

same time-base errors. The result was a stable 3.58-MHz signal suitable for recombining with the luminance. This heterodyne process is the same in principle as that used in the EIAJ and U-matic machines, so it will not be covered at length here.

THE DOWN-CONVERTED SUBCARRIER METHOD

The direct methods will work with machines that have a high writing speed, because these have an fm frequency high enough to accommodate the 3.58-MHz subcarrier. But the smaller machines, with their smaller head drums and lower writing speeds, cannot use an fm frequency high enough for the subcarrier. The only satisfactory way to record the color signal on these machines is to convert the subcarrier to a lower frequency.

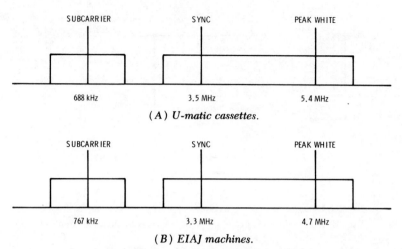

(A) U-matic cassettes.

(B) EIAJ machines.

Fig. 12-6. Color carrier frequencies on tape.

The method adopted is to separate the luminance and chrominance components with filters at the input of the machine. The luminance is treated as previously described, just as if it were a monochrome signal. The 3.58-MHz subcarrier is heterodyned down to a frequency below 1 MHz and then added to the fm signal. The fm signal serves as bias, just as in audio recording, and the combination of the fm signal and down-converted chrominance is recorded onto the tape.

The sidebands of the fm signal are such that the spectrum is fairly clear below about 1 MHz, and thus the lower subcarrier can be accommodated without trouble. Fig. 12-6 illustrates this and shows the lower frequencies used for both the EIAJ and U-matic systems.

On playback, filters are used to separate the low-frequency subcarrier from the fm signal, and each is fed into a separate circuit. The fm signal is treated as previously described and is demodulated to produce the luminance signal. The chrominance signal is heterodyned back up to 3.58 MHz, and then it is added to the luminance to form a normal NTSC signal. It is during the up-conversion on playback that the timing errors are removed from the color signal.

This method is used in the EIAJ-format machines, the U-matics, and other cassette systems, and it will be described in some detail. The frequencies given are those of the EIAJ system.

Record

The recording process is straightforward and simple (Fig. 12-7). The luminance is separated from the chrominance and fed to the

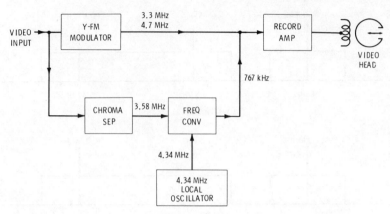

Fig. 12-7. Principle of recording with converted-subcarrier method.

fm modulator. The elimination of the color subcarrier from the luminance prevents interference in the color reproduction process. The fm signal is cut off at its low end by a high-pass filter to ensure that it does not occupy the spectrum below 1 MHz. The 3.58-MHz chrominance signal is frequency converted to 767 kHz by heterodyning it with the output of a 4.34-MHz crystal oscillator. This oscillator must be very stable to ensure interchange and good color with all machines. The resulting 767-kHz signal contains all the phase and amplitude information of the original 3.58-MHz signal, and it is now combined with the fm signal. The fm serves as a bias for the 767-kHz component, and to give a good signal-to-noise ratio in the color, the 767-kHz signal is set to be about 13.5 dB lower than the fm signal level.

The burst is separated from the incoming signal, and the burst level is used to control the signal level sent to the balanced modulator (Fig. 12-8). In the balanced modulator, the signal is heterodyned with the stable 4.34-MHz oscillator output. The difference frequency of 767 kHz is separated by the low-pass filter, is amplified, and is applied to the fm signal.

Playback

In playback, the head preamplifier amplifies the output from the video head and equalizes the signal to provide a flat response. Filters then route the luminance and the chrominance signal to their appropriate circuits.

The luminance signal is cut off at the low-frequency end and demodulated to produce the original luminance information. It is then delayed 0.7 microsecond (Fig. 12-9) to make sure it arrives at the output of the recorder at the same time as the color signal.

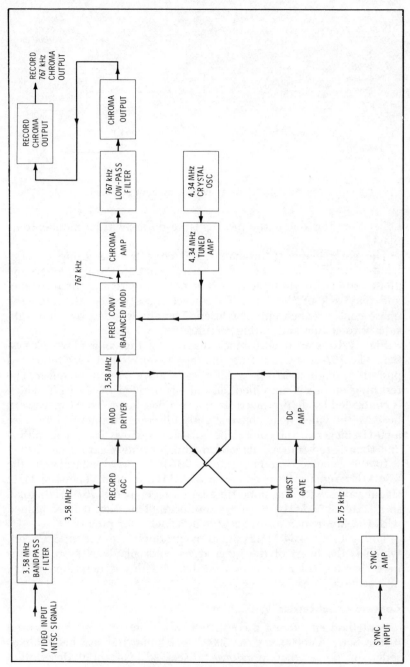

Fig. 12-8. Block diagram of recording process.

247

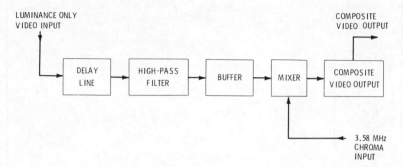

Fig. 12-9. Block diagram of luminance playback.

Other than this delay, the process is no different from monochrome playback.

The chrominance (chroma) signal consists of a group of sidebands around a 767-kHz carrier. These contain the frequency and phase errors caused by the time-base errors inherent in the recording and playback of the tape. To correct these errors, an automatic phase control (apc) circuit is used. The apc is aided by some subsidiary oscillator and gating circuits.

Fig. 12-10 is an overall block diagram of the color playback system. The 767-kHz signal from the tape is selected in a switcher and passed through a low-pass filter to remove any luminance. The resulting output is amplified, and agc is applied by an FET which is controlled by dc produced from a rectified burst. This is necessary because the video heads have slightly different outputs which cause a 60-Hz flicker in the color. The 767-kHz signal is further amplified and then is fed into a balanced modulator and a gating circuit.

In the balanced modulator, the 767-kHz signal mixes with the 4.34-MHz signal and produces a 3.58 MHz output. The 4.34-MHz signal varies with the same time-base errors as the 767-kHz signal, so a stable 3.58-MHz output is produced. This output is combined with the luminance to give a stable NTSC color picture.

The varying 4.34-MHz signal is produced in the apc circuit, which is the heart of the heterodyne color playback system. It is here that the color correction occurs, and this is described in the next section.

Converted Subcarrier With Pilot

A method employing a converted subcarrier with pilot was used in the Sony AV5000 and AV5000A machines, but has been superseded by the method previously discussed. This brief description is included because many of these machines were produced.

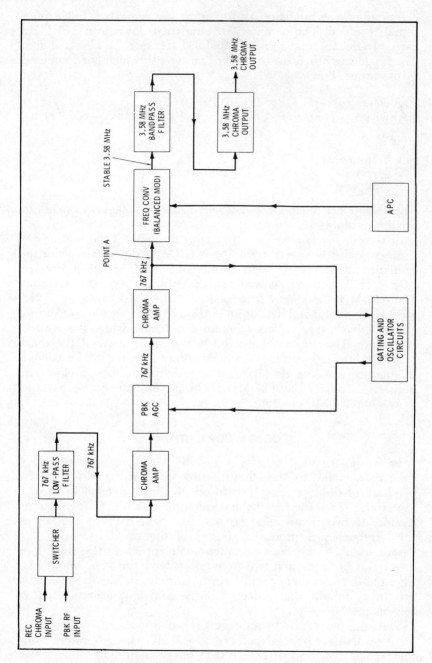

Fig. 12-10. Block diagram of color playback path.

249

As before, the luminance and chroma are separated, but a pilot signal is added. The chroma is heterodyned down to a 1.07-MHz carrier frequency, and a sine wave of 354 kHz or 357.9 kHz is added. The fm acts as a bias for the other two signals, which are combined in the ratios shown below.

AV5000: $Y + C + P = 1.5 + 0.19 + 0.05$ Pilot freq $= 354$ kHz
AV5000A: $Y + C + P = 1.5 + 0.35 + 0.08$ Pilot freq $= 357.9$ kHz

where,

Y = luminance,
C = chroma,
P = pilot.

The change of frequency and levels in the AV5000A was made to minimize color beats and noise.

In record, the luminance is put onto the tape in the normal fm manner, and it is filtered from about 1.7 MHz down. The color information is heterodyned with a stable 4.65-MHz signal to produce 1.07 MHz. The pilot signals are obtained from a separate oscillator. The 4.65-MHz frequency is formed from the third harmonic of the 358-kHz oscillator and the output of the 3.58-MHz crystal oscillator.

In playback, three filters separate the signals, which are treated separately. The pilot contains the time-base errors, and the third harmonic is mixed with the 3.58-MHz output from the local oscillator to produce a 4.65-MHz signal with time-base errors. This signal is used to convert the 1.07-MHz color information from the tape to a stable 3.58-MHz output.

COLOR CORRECTION

In the color-under, or down-converted, systems, the method of color correction is to sense and measure the time-base errors in the playback of the tape, and then to use this measurement to correct the errors. The basic idea behind color correction is simple heterodyning. If two frequencies are made to beat with each other, a third frequency is produced. If one of the two frequencies has variations in it, the same variations will appear in the third frequency. If both the first two frequencies have the same variations, then the third frequency will have no variations. This is covered a little more fully in the subsection on heterodyning near the end of this chapter.

The trick in vtr's is to produce the same variations in two frequencies that are heterodyned together. One frequency comes from the tape in the form of the chroma information and contains the time-base variations. The second frequency with the same time-base variations is produced by measuring or sampling the color burst or pilot signal from the tape.

Everything played back from the tape contains the same errors. These are in the chroma signal, the burst, and the pilot track if one is used. The burst can be separated from the color signal by a gate; the pilot is separate by its nature. If these are heterodyned with a stable oscillator, sum and difference frequencies are produced that contain the same time-base errors.

Simplified block diagrams of the converted-subcarrier method are shown in Fig. 12-11. Although details may vary, this is the basic idea of the EIAJ machines and the U-matic cassettes.

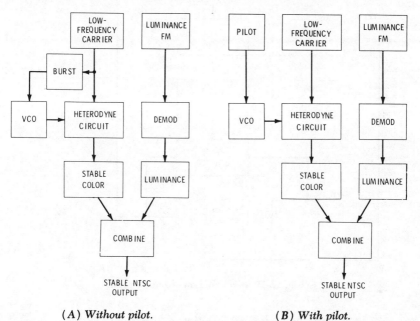

(A) Without pilot.　　　　　　(B) With pilot.

Fig. 12-11. Playback with color recorded separately.

The APC Circuit

The apc circuit is the most important part of the whole color playback system. The incoming 767-kHz signal is split into two paths at point A in Fig. 12-12. Path 1 carries the signal that will be heterodyned with the varying 4.34-MHz signal, and path 2 carries the 767-kHz signal that is used to produce the varying 4.34-MHz input. (Varying frequencies are indicated by an asterisk.)

The second path is an input to the apc loop (circuit). This input is heterodyned with a crystal-controlled 3.58-MHz signal to produce

251

a color signal of 4.34 MHz. This signal contains all the color video information and is unsuitable for use, but its burst is usable because it contains all the errors inherent in a particular line. The burst is separated to produce a pure sine wave of 4.34 MHz. This is not long enough in duration to heterodyne with the 767-kHz input on a continuous basis, so a continuous 4.34-MHz signal must be produced from this burst. The continuous signal is obtained from

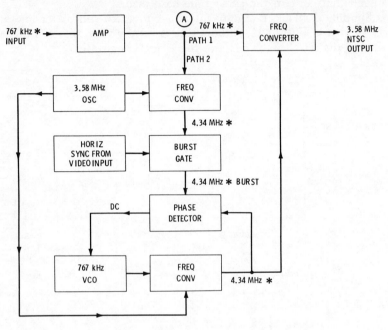

Fig. 12-12. Block diagram of apc circuit.

a 767-kHz voltage-controlled oscillator and the 3.58-MHz crystal oscillator. The result of this comparison is a 4.34-MHz signal that is frequency controlled by any variations in the vco. This 4.34-MHz signal is compared in phase to the burst to produce a dc voltage that is used to vary the output of the vco.

As the burst varies, so will the vco, and so will the 4.34-MHz frequency. This final 4.34-MHz signal is a continuous wave the frequency of which is controlled by the time-base errors from the tape on a line-by-line basis. This signal is fed to the balanced modulator in the main chain to heterodyne with the 767-kHz input from path 1. The result is a stable 3.58-MHz output. This output is amplified and then combined with the luminance signal, giving a stable NTSC picture at the output of the machine.

252

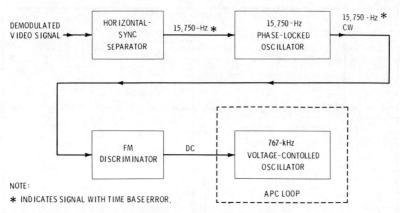

Fig. 12-13. Block diagram of afc circuit.

AFC Loop

In the slow-speed and still-frame playback modes, the relative head-to-tape speed is decreased by 2 percent. As a result, all the frequencies from the tape are also reduced by 2 percent. This is far too wide a change to be corrected in the apc loop, so to maintain a stable color picture a separate circuit is required. This is the automatic frequency control, or afc.

The 15,750-Hz horizontal-sync pulses from the tape are reduced 2 percent form their normal frequency, and they are subject to the same time-base errors as the 767-kHz signal from the tape. The horizontal-sync pulses are separated and then are detected by an fm discriminator (Fig. 12-13). This produces a dc output proportional to the time-base errors. This dc is fed to a second input of the 767-kHz vco in the apc loop. It slightly reduces the base frequency of the oscillator and then varies it with the time-base errors. This simple process ensures fairly good color in the slow-speed and still-frame modes.

U-Matic Cassettes

The method of color recording adopted for U-matic cassette machines is essentially the same as that described above, but the frequencies used are different. The color is recorded as sidebands of a suppressed 688-kHz carrier. Fig. 12-6 compares the frequencies of the two systems.

CIRCUITS FOR THE CONVERTED-SUBCARRIER METHOD

The next few paragraphs are a presentation of some of the more important circuit sections used in the converted-subcarrier color

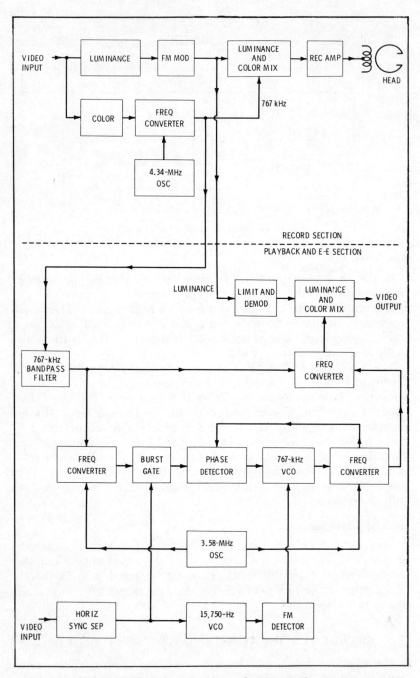

Fig. 12-14. Block diagram of complete EIAJ color system.

system. They are not explained in detail, but merely included to enable the reader to recognize them in a service manual and to give some idea of how the necessary functions are accomplished in the color correction process. These circuits are the balanced modulator, voltage-controlled oscillator, burst gate, phase detector, frequency converter, and color noise canceller. Fig. 12-14 is a fairly complete block diagram of the EIAJ color system.

The Balanced Modulator

The object of the balanced modulator is to accept two different frequencies at two input points and to produce an output that contains the sum and difference frequencies. These frequencies can be separated by filters after the modulator.

Fig. 12-15 is a diagram of a simple balanced modulator. The 4.34-MHz input is applied between the center taps of both transformers,

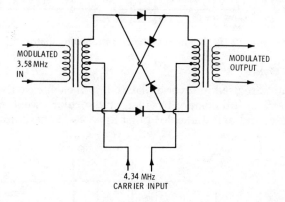

Fig. 12-15. Balanced modulator.

and the 3.58-MHz chroma is applied to one primary. The output is two sets of sidebands without a carrier. The lower set is centered at 767 kHz, and the upper set is centered at 7.92 MHz. A low-pass filter blocks the 7.92-MHz set and allows the 767-kHz set to pass to the following circuitry.

Voltage-Controlled Oscillator

The voltage-controlled oscillator can take many forms, and in fact the normal fm modulator used in a vtr is a vco. The form shown in Fig. 12-16 is a Colpitts oscillator in which frequency control is achieved by varying the dc level applied to a varicap. The varicap changes its capacitance as the dc changes and thus changes the frequency of the oscillator. The dc level is controlled by the output of the phase detector.

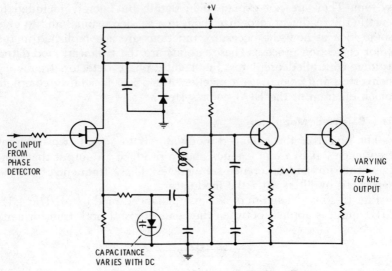

Fig. 12-16. Circuit of voltage-controlled oscillator.

The Burst Gate

Burst gates are used in several places, but in each case the purpose is the same. The full composite video or the color signal only is fed into the gate, and only the color burst appears at the output. In the

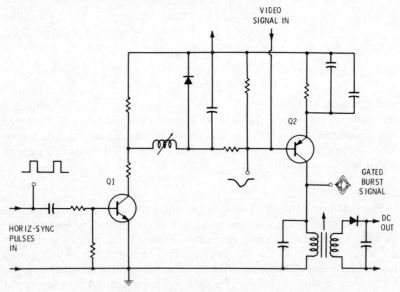

Fig. 12-17. Circuit of burst gate.

circuit of Fig. 12-17, separated horizontal-sync pulses are applied to Q1 and then to the base of Q2. The video signal is also applied to this same point. Transistor Q2 is normally in a nonconducting state, and the presence of the video signal is not enough to cause conduction. The arrival of the horizontal-sync pulse causes conduction. The inductor and capacitor delay the horizontal pulse slightly so that the conduction of Q2 occurs at the time of the burst signal in the video input. The result is that only the burst can appear at the output of Q2. These few cycles of burst are transformer coupled to a rectifier, the dc output of which is determined by the burst amplitude and is thus suitable for the control input to an agc circuit. The burst signal need not be rectified, but can be applied to a phase comparator for frequency-correction purposes.

Phase Detector

The object of a phase detector is to compare two signals at its inputs and produce an output—usually dc—that depends on the difference in phase of the two inputs. The two inputs need not be of the same frequency, but they often are. Fig 12-18 is a typical circuit.

Frequency Converter

The object of the frequency converter is the same as that of the balanced modulator, to accept two different input frequencies and

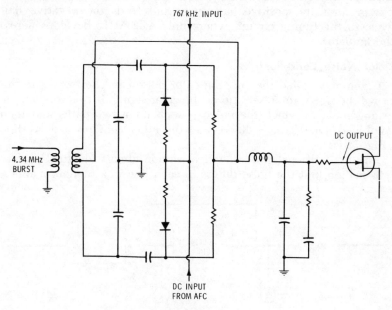

Fig. 12-18. Phase-detector circuit.

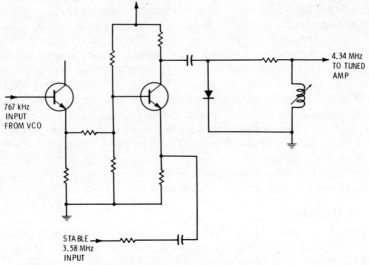

Fig. 12-19. Frequency-converter circuit.

produce an output frequency equal to their sum or difference. In the circuit of Fig. 12-19, 767 kHz and 3.58 MHz are mixed to produce 4.34 MHz. The amplifier is operated in its nonlinear region so that only the 4.34-MHz envelope is detected. The diode in the output grounds the positive half cycle and feeds the negative half cycle to the tuning circuit, where only 4.34 MHz develops across the inductor.

Color Noise Canceller

Commonly called the cnc, the color noise canceller serves to improve the signal-to-noise ratio of the color signal. It is a very simple arrangement in which the video is sent along two paths and then combined. One path is direct, and the other contains a delay line (Fig. 12-20).

Any two adjacent lines of video contain almost identical picture information, but the noise differs in each line in a random manner.

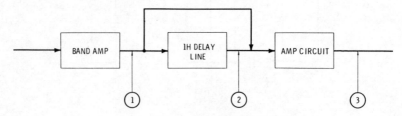

Fig. 12-20. Color noise canceller.

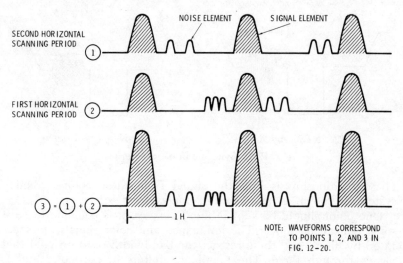

Fig. 12-21. Signals in color noise canceller.

Thus in Fig. 12-21 the two signals represented by waveforms 1 and 2 are the same in picture content but different in noise content. If these two signals are added, the picture elements become doubled in size, but the random noise elements do not. Thus the ratio of the signal to the noise is increased by a factor of 2, or 3 dB.

COLOR CORRECTION REVIEW

With all of the color-correction methods that have been described in this chapter, it should be clearly understood that only the color subcarrier is corrected and has its time-base errors removed. The result is that the subcarrier and burst are made sufficiently time stable to trigger the local oscillator in a monitor and produce a satisfactorily stable picture on the screen. In none of these methods is the luminance part of the signal corrected. Its stability is entirely determined by the servos of the machine. If the timing errors in the horizontal sync are about 1 microsecond or less, the signal is stable enough for most purposes. With the smaller helical machines, where much larger errors are seen, the stability of the picture on the screen is a function of the horizontal afc in the monitor.

Because the burst and subcarrier are now stable and the horizontal sync is not, the static phase relationship between them that existed in the original NTSC signal has been lost. This can be seen easily by viewing a horizontal-sync pulse and burst on an oscilloscope. An NTSC signal will resemble Fig. 12-22A, while the output from a small helical vtr will resemble Fig. 12-22B. Because this

259

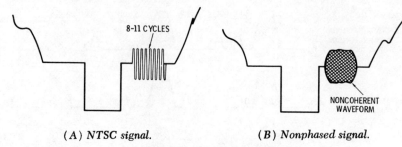

(A) NTSC signal.   (B) Nonphased signal.

Fig. 12-22. Sync and burst waveforms.

phase relationship is lost, the signal from these color-correction systems is known as *nonphased* or *noncoherent* color.

One point should be kept in mind because it is common to most small helical machines: The luminance and color signals are separated by filters, and these restrict the bandwidth of any signal that passes through them. The luminance suffers in particular and is often reduced to a bandwidth of no more than 2 MHz. This is acceptable only because the color masks the loss of detail and because many small tv sets have limited resolution.

A recording method that is not used is to keep the chroma information at 3.58 MHz and add this to the fm carrier during recording. A bias frequency should be about 3 to 5 times the highest frequency that it carries, and this would mean using about 15 MHz as the fm carrier frequency. Such a high carrier frequency would require a scanning speed around 1500 in/s, which is not practical in small helical machines.

GENERAL TOPICS RELATED TO COLOR

The following brief subsections deal with subjects pertinent to color in general and color vtr's. They are not intended to be complete or exhaustive, but merely to provide a background to some of the material in the chapter.

Heterodyning

Heterodyning is simply the mixing together of two different frequencies. This technique is well known in electronics and is used in many places, perhaps the most common being in radio and tv receivers.

When two different frequencies are mixed, they react to produce other frequencies. The two most important and pronounced of these new frequencies are the sum and difference of the two originals. If F1 and F2 are the two original frequencies and F1 is greater than F2, the new frequencies are:

260

$$F1 - F2 = F_d$$
$$F1 + F2 = F_s$$

Other, weaker frequencies are produced by further interactions of F1, F2, F_d, F_s, etc.

If F1 and F2 are constant, then obviously so are F_d and F_s. If F1 or F2 is varying, then F_d and F_s will contain the same variations.

$$(F1 + f) - F2 = F_d + f \qquad (\text{Eq. 12-1})$$
$$(F1 + f) + F2 = F_s + f \qquad (\text{Eq. 12-2})$$

If F2 varies the same as F1, then:

$$(F1 + f) - (F2 + f) = F_d \qquad (\text{Eq. 12-3})$$
$$(F1 + f) + (F2 + f) = F_s + 2f \qquad (\text{Eq. 12-4})$$

Now consider F2 varying the same amount as F1 but in the opposite direction. Then:

$$(F1 + f) - (F2 - f) = F_d + 2f \qquad (\text{Eq. 12-5})$$
$$(F1 + f) + (F2 - f) = F_s \qquad (\text{Eq. 12-6})$$

Of the above equations, Eqs. 12-3 and 12-6 are of interest in the color playback systems in vtr's. In these equations, F1 and F2 represent the two frequencies to be heterodyned, and f represents the time-base errors introduced by the tape playback.

Delay Lines

When an electronic signal passes through a circuit, it does not appear at the output at exactly the same instant that it was applied to the input. In many applications, this slowing down is insignificant, but in certain situations it assumes major proportions. A millionth of a second means little in our everyday lives, until we look at a tv screen. A 5° change of phase in the color signal can be noticed. This corresponds to a timing error of about 4 billionths of a second. Delays introduced into a signal by electronic components can reach nearly 1 microsecond, which is 250 times longer. Delays of 0.5 to 1 microsecond are quite common in helical vtr's, and such delays are more than enough to disrupt any color signal. To avoid this disruption, it is sometimes necessary to introduce delays into a signal path so that two signals can be made to arrive at a common point at the same time. This is required prior to recombining the luminance and the chroma information; the luminance is delayed typically 0.7 microsecond.

This artificial delay is caused by passing the signal through a delay line, which is merely a series of capacitors and inductors arranged so that any signal will be delayed by a given length of time. Physically,

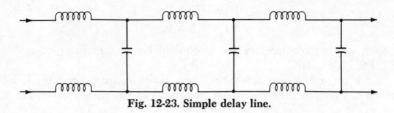

Fig. 12-23. Simple delay line.

these are quite small devices, and they are often mounted directly into printed circuit boards. Usually, they are manufactured with a given delay time and are not adjustable. Fig. 12-23 is a typical circuit of a delay line.

PAL

The PAL (*Phase Alternation Line*) system, which was invented in Germany several years ago, is a variation of the NTSC system. It has been adopted by several of the major European broadcasting organizations and has been followed by the nonbroadcast tv industry in those countries. It is essentially the same as NTSC, except for one major difference. The phase of the color signal is reversed on each alternate line. These reversals are averaged by the receiver circuits and cancel out many of the phase errors. This system requires very little modification of the basic system and introduces no new problems or major errors. Most manufacturers of vtr's produce models for use in this system.

SECAM

A French system called SECAM was the original European alternative to NTSC and would probably have gained universal European acceptance if PAL had not been introduced. It is not vastly different from the NTSC system, but it uses an fm method of carrying the color instead of the NTSC phase method. The color information is also carried sequentially instead of simultaneously. The main advantages of SECAM are a much simpler receiver design and complete freedom from the phase errors of NTSC. It has gained acceptance in France and the countries of Eastern Europe. The word SECAM means sequential with memory; it is not an acronym.

Color Test Patterns

Test patterns in tv are used for several purposes. In color work, the most common pattern is the color bars. These can originate from a card in front of a camera, but they are more often produced electronically.

Two distinct patterns are used. Fig. 12-24A represents a series of vertical bars in the top portion of the screen, and Fig. 12-24C shows

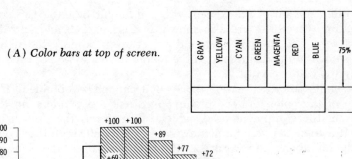

(A) Color bars at top of screen.

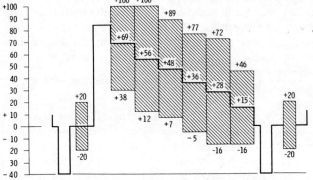

(B) Waveform of upper bars.

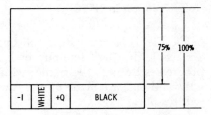

(C) Test bars at bottom of screen.

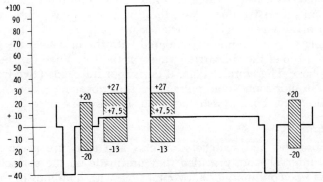

(D) Waveform of lower bars.

Fig. 12-24. Color test pattern.

the bottom section of the screen with a different pattern. The corresponding waveforms are also shown, in Figs. 12-24B and 12-24D, respectively. The equipment should be set up to produce the signal levels shown in the diagram, and the monitors should be adjusted to display the colors correctly.

In a nonbroadcast situation, perfection cannot be obtained. The quality of the color and the pictures is directly dependent on the quality of the equipment used, which often means the price paid for it. If a color-bar generator is available, about 30 seconds of color bars should always be recorded at the beginning of a tape. This allows the program playback to be set up correctly at a different location and on another machine, and it can often be invaluable in servicing if a machine should give trouble at some time.

Color Packs

Several of the early models of color machines did not have the color circuitry as an integral part of the machine; it was contained in a pack that was provided separately. This is an outmoded method, since the more modern techniques and the introduction of integrated circuits allow easier incorporation into the main chassis of the machine. Therefore, separate packs are not covered in this book.

Color Framing in NTSC

A monochrome picture consists of a frame that is divided into two fields. Field 1 starts with a full line and ends with a half line; field 2 starts with a half line and ends with a full line. Horizontal sync progresses through the frames with no interruptions. If monochrome edits and cuts are made on the correct field, no picture disturbances will result.

In color, there are $227\frac{1}{2}$ complete cycles of subcarrier in one horizontal line. As a result, there is a $180°$ phase difference in the subcarrier on successive lines. Also, there is a $180°$ phase difference in the subcarrier when successive frames are compared; i.e., if one field beginning with a full line is compared with the next field beginning with a full line, the subcarrier waves in the two lines will be $180°$ out of phase. Therefore, in color it takes four full fields (two frames) before the horizontal-sync pulses and the subcarrier phases match. Thus the NTSC color system has a four-field cycle.

Color Editing

In most vtr's and switchers, all cuts and edits are made in the vertical interval, and, provided they are made on the correct field, there will be continuity of horizontal sync with no jumps in the picture. In color, if the edits and cuts are made in the vertical interval on a line where there is no burst, the same field matching will suffice. This is because the local oscillator in a monitor has plenty of time to relock to the burst in the last few lines of vertical blanking and be stable by the time the active lines start. So, edits are quite satisfactory on a two-field basis.

This situation changed when the digital time-base corrector was introduced. To see why, assume the last line of a frame is stored in the memory. An edit is now made in the vertical interval, and the color frame is wrong. When the new line is put into the memory, instead of its subcarrier phase being 180° different, it is the same as that of the previous line.

When the signal is read out from the memory, a clock controlled by the station sync is used, and its subcarrier will be alternating correctly. The stable incoming sync and subcarrier are put onto the output signal from the memory as horizontal sync and burst, and the burst phase alternates correctly with each line. But suddenly the burst is matched with a subcarrier that is 180° out of phase, and the color will change radically—all the vectors will be shifted 180°.

To correct for this, the time-base corrector alters the memory readout. It starts reading out the line at a point corresponding to one-half subcarrier cycle from the left edge of the picture. However, the display of this signal on the picture tube still starts at the left edge, and the result is that the line jumps a little to the left. Unfortunately, a jump of half a subcarrier cycle (140 nanoseconds) is visible. To prevent this, later vtr machines all have a color-framing circuit in the capstan servo, and they place a color-framing pulse on the control track. This ensures that all edits are made only when the color frames are properly aligned.

Color Edits in Small Helical Machines

In the small inexpensive helical machines, color editing is unsatisfactory at best. Usually, the capstan and head servo will effect an alignment that allows the edit to be made in the vertical interval, and it can appear quite stable on the screen. But the large time-base errors and lack of color framing make color editing almost impossible. Color editing on these machines will sometimes appear satisfactory on the screen, but this is due to the quick recovery time of the monitors and is not due to the circuits in the vtr's.

Whether a color edit is acceptable or not is a matter of personal choice. It is often dependent on the type of studio or the type of material. Once a bad edit is on a tape, it is there until the tape is re-recorded.

13

Time-Base Errors and
Their Correction

The helical vtr has given excellent service in industry and education for many years, but only recently has it found significant application in broadcasting and teleproduction. The main reason why it was not used initially in tv studios was that its sync timing on playback contains excessive timing errors when compared to the stable signal from the station sync generator. These timing errors are called *time-base errors*. The apparent stability of the helical picture when viewed on certain tv monitors is due to the fact that the horizontal-oscillator circuits of these monitors have been modified. These pictures are no criterion of stability. (It should be realized that all video tape recorders, even the quad machines, produce time-base errors. However, the errors in quad machines are small and can be corrected easily to make the output stable enough for broadcasting.)

Before any vtr can be used in broadcasting or serious teleproduction, the time-base errors must be corrected. This is necessary so that the playback signal from the vtr will be stable at all times and can be used as a source of video in production. It is desirable to use prerecorded material in many productions. For example, scenes shot in a studio may be interspersed with those shot on location, and different scenes may be taped on different days. Thus it is necessary for tapes to be played back and mixed with other tapes and with live camera signals. In order for this to be possible, the signal played back from the vtr must be as stable as that from the studio cameras.

In general, the signal from a vtr is *not* as stable as that from a camera. It contains large variations in the timing of its sync pulses and subcarrier that make it almost unusable. Also, its sync and sub-carrier will not be aligned in time, or phased, with those of the cameras. Therefore, the object of time-base correction is to remove the timing errors and to phase the signal to the studio sync generator. Then the vtr can be used as an extra source of video in all productions.

THE MAIN CAUSES OF TIME-BASE ERRORS

Timing errors from a vtr are visible only during playback, but they can be introduced during both the record and playback processes. There are four main causes of timing errors:

1. Tape-motion irregularities
2. Head-motion irregularities
3. Tape-dimension changes
4. Skew errors

These together cause variations in the scanning speed and in the linear speed of the tape in both record and playback. The problems can be compounded when the errors are different on playback and record. In addition to the problems described below, all errors produce horizontal instability of the picture and cause wrong colors, lack of color, or unstable color on the screen.

Tape-Motion Problems

Slight tape-speed instabilities, similar to flutter and wow in an audio machine, cause errors in the sync and subcarrier timing. A high-speed flutter appears as vibrations running through the picture and a blurring of vertical lines. The slower wow effect causes wobbling of the raster from side to side. In audio, slow or fast tape speed causes pitch variations. In video, many instabilities are introduced into the picture, and the tape may be completely unplayable.

Head-Motion Problems

Slight instabilities of the head rotation produce effects similar to those described above. Another problem is caused by head wear during normal use. As the head wears, the penetration becomes less, and the minor local tape stretch at the point of head contact is altered. This change is large enough to cause timing errors that need to be corrected.

Tape-Dimension Problems

Problems involving tape dimensions arise from several different causes. Temperature and humidity changes cause minor expansion and contraction of the tape backing. Temperature variations also cause minor changes in the dimensions of the metal parts of the machine, which will affect the tape path and produce the same result as a stretched tape.

Tape-dimension errors are minimal in quad machines because tape tends to stretch in the direction of its length and not across its width. Hence, the track-length changes due to tape changes are almost negligible in the quad format. In the helical format, however, tape stretch is in the same direction as the video tracks, and so it becomes a major problem.

Skew Errors

Skew errors occur when a scanning head leaves one side of the tape and the same or another head meets the other side of the tape. In helical machines, these two points are separated along the tape by a large distance, and tape stretch introduces large timing errors. This is common in helical machines and appears as an offset and flagging in the picture at the point where the head changeover occurs. Skew problems can be alleviated by correct tension control and by ensuring that the changeover point is just before or within the vertical interval.

Overall Effects

All of the above errors occur at random and in combination, and collectively they are large enough to ruin the stability of the signal when it is recorded and played back. These errors are found in their least degree in the quad machines and are at their worst in the small inexpensive helical machines. To produce a stable picture from any vtr, these timing errors must be corrected, and for this a time-base corrector is used.

QUAD TIMING PROBLEMS

It is convenient to look first at the timing problems in the quad machine and see how they were solved. Because most of the ideas for timing-error correction were applied first to quad machines, studying these applications helps to put the methods in perspective and to relate them to helical machines.

The early quad machines were monochrome only. Their servos were referenced to the stable station sync generator, which also permitted phasing of the sync pulses and thus allowed the machine to be used as a video source in production. The servo circuits controlled the head and capstan to a fine degree and could produce timing errors as small as about 1 microsecond peak-to-peak in the horizontal-sync pulses. This is acceptable for monochrome work because the slight horizontal shift is barely visible on a monitor, the output can be mixed with other sources, and the error will not upset another vtr during copying.

Although this timing error is acceptable for monochrome, it is far too great for color. A jitter of less than 10 nanoseconds can be seen, and to produce stable color the timing errors must be less than this. A stability of this order is far beyond the capability of electro-mechanical servos, and purely electronic means must be used to obtain a stable color signal.

In attempting to produce stable color from quad machines, three basic methods were tried:

1. Heterodyned color
2. The demod-remod method
3. Electronically variable delay lines

The first two methods were tried by different manufacturers in their quad machines, and they were also applied to some early helical machines. These have been covered in Chapter 12. With these two methods, it is important to realize that they corrected the color timing errors only and made no attempt to improve the horizontal stability of the machine. Because of this, the relationship between the horizontal-sync pulses and the subcarrier phase was lost, and these methods were called *nonphased* methods. Although they produced stable color on a monitor, they were not perfect in terms of their overall signal performance. Only the third method provides a true correction for the whole signal.

The EVDL

The electronically variable delay line (EVDL) is a very successful method of producing stable color from a vtr. The idea is to pass the signal through a delay line and to compensate for the errors by altering the delay time slightly.

The principle of operation of the EVDL is shown in Fig. 13-1. The varactor control voltage is an error signal produced by comparing playback sync with station sync. This error voltage alters the capaci-

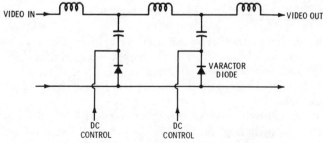

Fig. 13-1. Electronically variable delay line.

tance value slightly and thus alters the effective length of the line. By comparing successive playback pulses, another error voltage can be produced to remove velocity errors in each horizontal line.

Delay lines of this sort work satisfactorily only for errors of less than 5 microseconds. Lines with a longer delay seriously affect picture quality and cannot be used. The effective length can be varied about 20%, and so errors of 1 microsecond can be reduced to about 30 nanoseconds. This is excellent for monochrome, but it is still not good enough for color. However, the signal is now in a range where the color timing errors can be corrected.

To correct the color errors, another delay line is used. In this case, the error voltages are developed from comparisons of burst phase. The line is much shorter than the first one, and color errors can be reduced to about 10 nanoseconds or less.

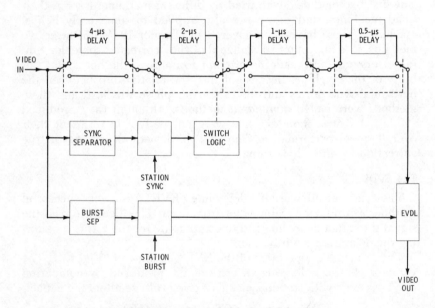

Fig. 13-2. Use of switchable fixed delay lines.

An alternative method is to use fixed delay lines that are switched in and out of the circuit. Each line has a binary relationship to the next one, and any required combination can be switched into the signal path (Fig. 13-2). The shorter EVDL for color is placed at the end of this chain.

Both of these methods have been used successfully with quad vtr's and high-quality helical machines. They cannot be used with the smaller helical machines because the raw time-base errors are too great. The length of line needed would degrade the picture excessively.

Other Delay Devices

Two recent advances in technology have also produced successful time-base–correction methods. The first is the *acoustic wave device*. To use this device, the video signal is converted by a transducer into a wave that physically travels along the surface of a semiconductor. The wave travels at acoustic, not electric, speed. At the other end of the device, another transducer reproduces the video signal. The result is that the small physical length presents an enormous time delay without severe signal degradation. Variable delays can be produced by switching in devices of fixed delay time. This type of delay device is not in common use, but it has been incorporated into at least one stand-alone time-base corrector.

The other new method is to pass the video signal through a *charge-coupled device* (CCD). Here, the signal is passed "bucket-brigade" fashion down a line of semiconductor capacitors. The longer the line of capacitors, the longer the delay. A variation of this is known as the *serial analog memory* (SAM). The CCD and the SAM have not yet appeared in commercial time-base correctors, but both are finding applications in other areas of video and audio. The CCD has been used as a solid-state camera, and the SAM has been used as an audio delay device when it was desired to produce echo and other effects.

The methods described in this section are known collectively as *analog* correction methods.

HELICAL TIME-BASE ERRORS

The newer high-quality helical machines can produce monochrome pictures that are comparable to those from a quad machine and that are stable enough for broadcasting. The early machines of this type, such as the Sony MV 10,000, used long digital switched delay lines and a shorter EVDL to produce a stable color output. Although these machines received minimal broadcast use, they found use in serious industrial applications and in some teleproduction houses that specialized in the making of multiple copies of programs on video cassettes.

The servos in a medium-quality helical machine can produce an output signal with as little as 5–20 microseconds of error, and a small inexpensive machine produces pictures with errors that could be several lines in length. Such errors are beyond the capability of the delay-line correction methods, and thus these machines cannot be used for high-quality broadcast-type work.

The instability of the helical picture comes mainly from the very long track, but other factors such as dimensional changes in the tape and head penetration are also heavy contributors. An example will make this clear. Suppose a 15-inch track becomes lengthened by 0.001%. This is 150 microinches. Calculations show that one horizontal line takes up 57,140 microinches and 360° of subcarrier takes up 252 microinches. If a 5° change can be seen, this means that 3.5 microinches of change will produce a visible hue shift. A 150-microinch change in track length is equivalent to about 214° of subcarrier. So, it is obvious that the timing errors in helical machines are more than enough to ruin the color completely, and only in the best machines are they small enough for monochrome playback.

Two early methods of color correction were tried with helical machines and found some success in producing stable pictures on a monitor. The servos achieved a timing stability that would produce a stable raster on a modified tv-monitor. The color was then separately corrected by the heterodyne or demod-remod technique to the point where a monitor local oscillator would lock and produce stable color. However, when the waveforms are viewed on an oscilloscope, the timing errors compared to station sync are enormous, and mixing with other signals, timing through a switcher, or multiple-generation copies are all out of the question. For nonbroadcast purposes, this type of playback signal is quite satisfactory, but for broadcast-type work it is useless.

THE DIGITAL TIME-BASE CORRECTOR

The digital time-base corrector, usually referred to simply as a *TBC,* was the first successful product to grow out of digitizing the video signal. Although other factors such as high-energy tape and improved mechanical engineering helped, it was the TBC that transformed the helical vtr into a broadcasting machine. The TBC can correct very wide timing errors and produce a broadcast-quality picture at its output when used in a "stand-alone" configuration between the output of the vtr and the input to a studio.

Basic Operation

The principle on which the TBC works is shown in the block diagrams of Fig. 13-3. The video signal is digitized by the methods described in Chapter 17. Fig. 13-4 shows the part of the signal that is digitized.

The sampling frequency is controlled by the horizontal sync and subcarrier from the tape, so it contains the same time-base errors as the signal. These are stable compared to each other, so the sampling is conducted at regular intervals within the signal. The resulting bit

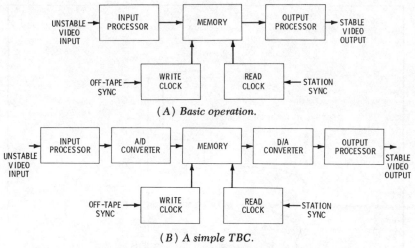

(A) *Basic operation.*

(B) *A simple TBC.*

Fig. 13-3. Principle of time-base corrector.

words are then entered into a computer-type digital memory by a clock that is also controlled by the off-tape signal. Thus, although the words are entered into the memory at a varying rate, they are entered in an orderly manner with correct incrementing of the memory addresses.

Once the signal is in memory, it can remain there indefinitely without degradation. The words can be "marched out" at a slightly later time and at a rate determined by the very stable station sync and subcarrier. The words are then transformed back into a video signal at a rate controlled by the station sync so that the signal has no timing errors.

The size of error this method can correct is limited only by the size of the memory used. Early machines had about three lines of storage, but later models exceed this greatly. The frame store, for example, stores a complete tv frame in a digital memory. This is basically a large-range TBC that finds use in areas other than with vtr's. Note that the correction range is not the same as memory size; for example, a TBC with three lines of memory may have a correction range of only $\pm\frac{1}{2}H$.

There are three main types of TBC in use. They are listed here in ascending order of complexity:

1. Picture straighteners. These are the simplest type and make no attempt to produce a broadcast-quality picture. They merely stabilize the signal so that the picture on a monitor or tv set is acceptable, with stable color and no hooking or skew. They are ideal for educational institutions or small catv systems.

273

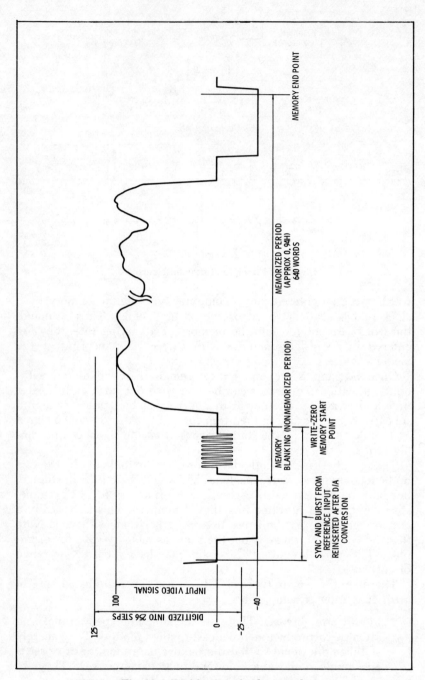

Fig. 13-4. Digitization of video signal.

2. Averaging TBCs. These use a phase-locked loop (PLL) at the input with a frequency close to the subcarrier, and they are pulsed by the incoming sync or burst to form the write clock. This is similar to genlocking to an incoming video signal. This type averages errors over several lines and works very well on slow and medium-rate errors. Often, this type will ignore a large error and clean up a bad signal fairly well. In many respects, these are the best type to use with a small-format helical machine that has large time-base errors.

3. Line-by-line correctors. These make an instant correction as each new sync pulse and burst arrive. They can correct high-rate errors and bad edits. All quad EVDL time-base correctors are the line-by-line analog type. Some high-quality broadcast TBCs use a mixture of types 2 and 3.

Additional Functions

Fig. 13-5 is a simplified block diagram of a high-quality broadcast TBC, and it can be seen that several functions in addition to time base correction have been included. Since the signal is held in a memory, it is easy to perform various other corrections, and typical high-quality TBCs contain circuits for these. The most common extra facilities are:

Dropout compensation
Velocity compensation
Heterodyne operation
System phasing controls
Picture enhancement
Gyro corrector
Stable picture in shuttle modes
Advanced vertical-sync feed

These are covered briefly in the next few paragraphs.

Dropout Compensation—Dropout compensation is very easy in a digital system, and it affords high-quality performance. A digital dropout compensator (*DOC*) can work by reading the binary words out of memory again or by passing them through a chain of flip-flops; thus no loss of video quality occurs. By having the signal from the vtr switch to the repeated signal only when a dropout occurs, only the part of the line with missing video is affected. The DOC can be operated by sensing the rf level out of the vtr, as with the stand-alone analog DOCs. However, the modern broadcast helical machines contain the rf-dropout sensing circuit and deliver a TTL-level pulse at an output socket. This pulse is fed to the TBC to operate its DOC.

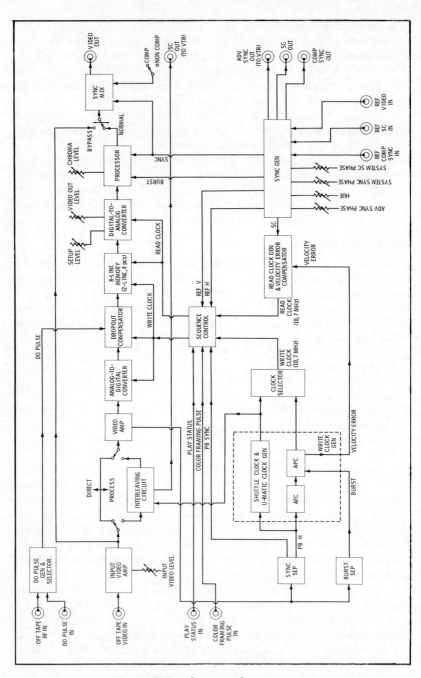

Fig. 13-5. Broadcast time-base corrector.

276

Velocity Compensation—Velocity errors show up as a hue shift from the left to the right side of the picture. They are caused by slight changes in the effective scanning speed of the head. The averaging type of TBC effectively selects an average of the error over a few lines and tends to reduce all the errors to that average, thus covering up the velocity errors in the general correction process.

In the line-by-line method, all the horizontal lines are corrected at their beginning only, and the errors that grow as the line progresses—i.e., the velocity errors—are not corrected. To correct the velocity errors, an assumption is made. Because of the short distance occupied by one horizontal line on the tape, the velocity changes are assumed to be constant over the line. Now the variations in burst phase of successive lines are compared, and an error voltage is developed from these comparisons. This voltage appears as a positive- or negative-going ramp, as in Fig. 13-6. It lasts for one horizontal line, and its height depends on the magnitude of the error. This ramp is used to vary the timing of the clock pulses, which corrects for the velocity errors and causes the digital samples to be read out at a stable rate.

Heterodyne Signals—In the color-under systems, the relationship between sync and burst is lost. In the TBC, sync and burst are not always digitized and stored with the luminance and chrominance components of the signal, but are added fresh in the output processing circuits. This means the phase of the burst can be adjusted to that of the subcarrier, thus restoring the phase relationship and re-creating a "coherent" signal.

Fig. 13-6. Velocity-compensation waveform.

In many situations, this requires a "two-wire" approach. A subcarrier phased with incoming sync is fed back into the vtr to control the oscillators in the color-recovery circuits. This usually requires a modification to the vtr, especially if it is an older type. This is called the *process* mode of operation. The more expensive TBCs use this process mode and also the *direct* mode, in which the subcarrier is not fed back to the vtr.

System Phasing Controls—In order to play back the vtr through a studio switcher and mix its signal with those of other vtr's and cameras, its sync and subcarrier must be properly timed or phased with the other signals. The sync generator in the TBC includes controls for this purpose.

Picture Enhancement—The enhancement of the video signal by digital means has been shown to be very effective, and some TBCs contain circuits for this purpose. Two typical corrections applied to the signal are:

1. Many small vtr's have chrominance-to-luminance delays that appear as a red smear in the picture. The enhancer separates the luminance from the chrominance, and a control is used to adjust their relative phase.
2. Many low-quality helical vtr's have a rapid falloff in the high frequencies (limited video bandwidth). The enhancer can provide a peak in the video response to improve the reproduction of high frequencies. Either a comb filter or coring is used to improve the signal-to-noise ratio by about 3 dB.

Care must be used with these techniques, because they can produce excessive video-signal spikes that will upset the TBC, and they can cause degradation of a high-quality signal.

Gyro Errors—Due to the movement of a portable vtr when in use, the head drum is subjected to irregular forces and stresses that can cause wildly uneven rotations during recording. These irregularities are recorded onto the tape and show up as extreme time-base errors on playback. They are known as *gyro errors*.

To lessen the effect of gyro errors, many high-quality portable machines have an FG coil or extra PG coils and vanes in the head drum to provide inputs to the head-drum speed servo. These inputs reduce the gyro errors but cannot eliminate them, and they may still be severe enough to cause trouble even when a TBC is used.

A TBC may not correct these errors fully because of a peculiar characteristic of how the digitized signal is put into the memory. The clock pulses that load the digitized signal into the memory usually are developed from the incoming sync pulses. Because of the circuit delays, it usually works out that when a line is loaded into memory the clock pulses loading it are generated from the sync pulse of the previous horizontal line. Normally this is not serious, because the timing errors from one line to another are very small. But with gyro errors, this difference can be enough to prevent full correction.

The manner in which this problem can be avoided is to delay the incoming signal by one horizontal line. This is done by passing it through a glass delay line, as in Fig. 13-7. The video signal is modulated onto an rf carrier, usually between 12 and 30 MHz, and this rf is delayed by the glass line. The video is then demodulated and applied to the A/D converter. The added circuitry for this process is called a *gyro adapter*.

Stable Pictures in Shuttle Modes—Because the digitizing of the incoming signal is controlled by the playback sync and subcarrier, the signal can be written into the memory in the fast-forward and rewind modes just as easily as it can be in the playback modes. Once

278

in memory, the signal can still be read out by the station sync pulses. The different read-in and read-out rates produce a stabilized picture with a speeded-up motion, similar in appearance to a comic film made with "stop-motion" photography. The color can be rendered stable up to about 7 times normal speed, and the monochrome up to about 30 times normal speed. The advantage of this is that it allows the operator to shuttle the tape back and forth at high speeds while searching for edits, etc., and this vastly reduces post-production

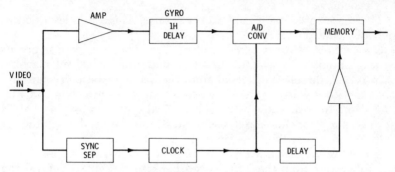

Fig. 13-7. Gyro-adapter function in TBC.

editing time. The principle on which this mode of operation relies is that not every line or frame has to be entered into the memory, and that a memory can be read at a very different rate from that at which the information was entered into it.

Advanced Vertical—When a vtr signal is played through a digital time-base corrector, the signal at the TBC output has been delayed by its time in the memory. The only way this final output signal can be made to coincide with the other signals in a tv system is for the playback signal from the vtr to be slightly advanced in time with respect to these other signals.

This time advance is achieved by feeding the vtr a vertical-sync pulse that is a few lines earlier than the pulse from the station sync generator. The vtr locks to this pulse so that its output is earlier than it would normally be. The advanced vertical pulse is generated in the TBC by counting horizontal lines from the previous station vertical-sync pulse.

The time by which the advanced vertical pulse is earlier than the station sync can be adjusted by means of a control on the TBC until the vertical intervals of the vtr signal coincide with those of the studio cameras when viewed at the output of the switcher. Usually this advance is about 3 or 4 lines.

Most digital TBCs have a coincidence circuit to compare the advanced vertical-sync output with station sync. This circuit drives three indicator lights, one for insufficient advance, one for too much advance, and one for correct advance.

TBC Operation

Operation of the TBC is simple. It is merely a matter of making the correct connections to the vtr and then setting the few controls on the front panel. Because digital circuits are used throughout, simple indicator lights can be provided to show when the correct setting is achieved.

There are three main modes of TBC use. These are described in the following paragraphs.

Connections With a Direct Color VTR—This is a typical broadcasting situation in which a quad or high-quality helical machine is in use. A direct color signal from the vtr is to be corrected and phased with a studio system. Typical connections are shown in Fig. 13-8. Note the use of the advanced vertical sync into the vtr. This is not necessary for simple playback, but it must be used when the vtr is to be phased with other inputs to a switcher.

A Heterodyne Color VTR in the Direct Mode—In this connection, the subcarrier from the TBC is fed back to the vtr to control the heterodyne recovery process when the tape is played back. This is a typical broadcast situation in which a high-grade helical machine is being used. Often the vtr must be internally modified by removing

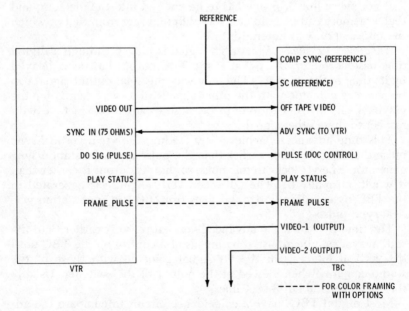

Fig. 13-8. TBC connections with direct color vtr.

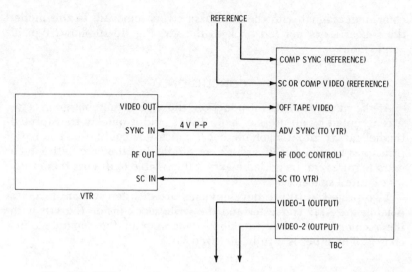

Fig. 13-9. TBC connections with heterodyne color vtr, direct mode.

its 3.58-MHz crystal in the heterodyne oscillator. The oscillator is then controlled by the incoming subcarrier. The output from the TBC will be a correctly phased coherent NTSC color signal. Fig. 13-9 shows the connections.

A *Heterodyne Color VTR in the Process Mode*—In this mode, the video output from the vtr is time-base corrected, but the color is not phased or coherent with the luminance. It will appear as a typical

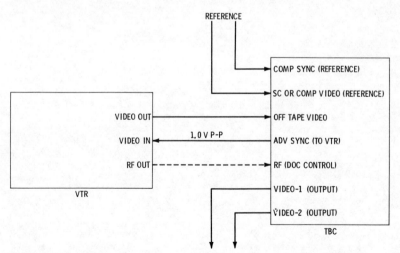

Fig. 13-10. TBC connections with heterodyne color vtr, process mode.

color-under signal with the time-base errors removed. In this mode, the subcarrier is not fed back to the vtr. Fig. 13-10 shows typical connections.

CONCLUSION

At the moment of its introduction, the digital time-base corrector was accepted by all users of helical vtr's, and it quickly transformed the helical vtr into a machine that broadcasters could use. The basic principles by which TBCs work are simple, but the ease with which extra features can be added makes it impossible to discuss them fully in a limited space.

Two projected future developments are of interest. One involves placing the A/D converter and the write clock inside the vtr, with the memory and other functions remaining in the separate TBC chassis. The other is a fully digitized vtr.

14

Video Cassette Machines

In addition to the open-reel machines discussed in the preceding chapters, the vtr field now includes cartridge and cassette machines. In these machines, the tape is contained in a protective enclosure, and the entire enclosure is inserted into the machine, where the tape is threaded automatically. The methods by which this is accomplished are described in this chapter.

Video cassettes have developed in two ways since their introduction. In the industrial and educational fields, the U-matic has become dominant, and a version with the improved Type 2 mechanism has become extensively used in broadcasting. In the consumer area, smaller machines such as the Betamax have been introduced with a considerable degree of public acceptance.

Several other types were introduced at about the same time as the U-matic format, but most have disappeared. The main exception to this is the Philips video-cassette machine, which is used extensively in Europe.

Because most of the cassette vtr's work on similar principles, this chapter will concentrate mainly on the important features of the U-matic machines. Also, the significant differences found in the Betamax machines will be introduced.

CASSETTE MACHINES

In essence, a video cassette machine is a vtr with a specialized mechanical construction and format. It is the same as an open-reel vtr in principle, and it differs mainly in appearance and operation. The major differences are that the tape is in an enclosed box or cassette, the threading is automatic, control circuits are included for this and other functions, and a tv tuner and an rf modulator are built in (this is to ensure acceptance as a home consumer unit). Fig. 14-1 shows a U-matic cassette player.

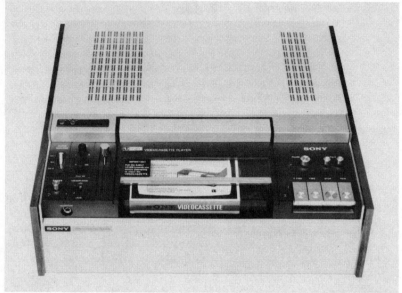

Fig. 14-1. Video cassette player.

Most of the manufacturers make two types of machines, a playback-only model and a record-playback model. The latter is the same in all respects as the former, except for the addition of the recording electronics and a tv tuner to give off-the-air capability.

THE CASSETTE

The cassette measures 8¾ in × 5½ in × 1¼ in, weighs 1¼ pounds, and contains tape for 60 minutes of recording at 3¾ in/s. The tape has a high-density chromium-dioxide or cobalt-doped ferric-oxide coating. Transparent leaders at each end of the tape attach it permanently to the hubs of open-faced reels inside the cassette. An opening along one edge of the cassette makes the tape accessible for pulling into the machine. This opening is guarded by a spring-loaded flap. To load the cassette into the machine, all that is necessary is to push it into the appropriate opening and let the machine do the rest.

To prevent erasure of a prerecorded tape, a plastic button is removed from a hole in the bottom of the cassette. This allows a small pin to enter, and the position of the pin serves to disable the record function.

THE MECHANICS OF CASSETTE MACHINES

In appearance, cassette machines are quite unlike any other tape machine in that the reels, tape, heads, etc., are not visible or accessible. Only the push buttons and operator indicator lights are visible and available on the front panel.

The cassette is loaded into the machine through a special slot in the front. When the cassette is pushed in fully, the slot automatically drops to a lower position where the automatic threading and loading can begin.

Just as in any other helical machine, the tape must describe a helical path around a rotating head drum. This drum is mounted within the machine at a level below that of the lowered cassette. Threading the tape around this head is accomplished by a large threading ring that rotates around the outside of the head drum and the tape guides.

When the cassette is inserted, the tape door opens to allow a threading roller to be placed in the triangular notch. This roller is mounted on the end of a pull-out arm, which is attached to the threading ring. Two types of threading ring are used, depending on whether the Type 1 or Type 2 mechanism is used.

Type 1 Transport

The Type 1 transport or mechanism was introduced with the first U-matic machines and is used in the smaller industrial and educational models. The threading ring is driven clockwise in the loading operation and counterclockwise in the unthreading operation. It is rim driven by a dc motor and a rubber-tired wheel, which is powered when a function is selected. The basic layout is illustrated in Fig. 14-2.

The tape is pulled out of the cassette and then around a wide-diameter circle (Fig. 14-3). The threading ring is not horizontal, but is placed on a slant. Tape guides are mounted on it as well as on the fixed part of the chassis, and these guide the tape in its downward circular path. The capstan pressure roller is mounted on the ring and is placed so that it follows the tape inside the loop. Eventually, the ring stops, and the tape and roller are in the positions shown in Fig. 14-4. Note the position of the erase, control, and audio heads in relation to the head drum and the path followed by the fully threaded tape.

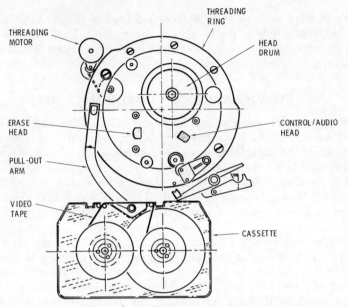

Fig. 14-2. Threading ring and heads.

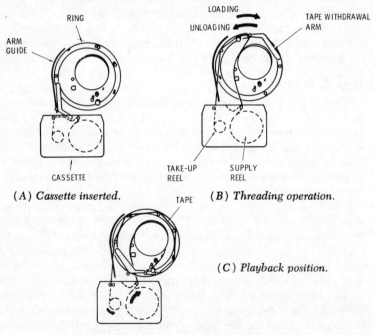

(A) *Cassette inserted.*

(B) *Threading operation.*

(C) *Playback position.*

Fig. 14-3. Stages of tape threading.

When the tape is fully wrapped and in position, it closes a switch that applies power to a solenoid to pull in the pressure roller and power the motors. The tape begins to traverse its operational path, and the head drum begins to rotate. From the operation of the start button to this point takes about 6 seconds.

The tape is pulled from the supply side of the cassette and past the guides and heads at a speed of 3¾ in/s. It is pulled back into the take-up side of the cassette by the tape-up spool, which is belt driven from the drive motor. The heads rotate at 1800 rpm and give an effective head-to-tape speed of 1000 in/s. This continues until the stop button is pressed or until the tape runs out.

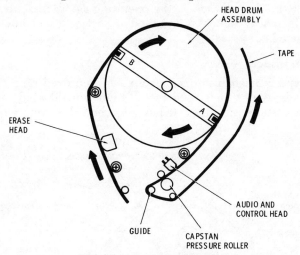

Fig. 14-4. Tape fully threaded.

When the stop button is pressed, the pinch roller is released, and the tape begins its retraction into the cassette. The threading ring revolves in a counterclockwise direction to feed the tape loop back to the cassette, and the take-up reel winds the tape back into the cassette. The supply reel is braked during the process. When the tape is fully inside the cassette, the ring stops, and a length of tape equivalent to about 10 seconds of program material is then rewound onto the supply reel. Throughout this operation, a yellow warning light on the front panel is illuminated, indicating that no further operations are to be attempted until the light goes out. The retraction operation takes about 6 seconds to complete.

The rewind and fast-forward modes are carried out with the tape entirely within the cassette. If it is required to rewind and play back a section just viewed, the stop button is pressed, and after the tape is fully retracted, the rewind can be initiated. The same is true for fast forward. Once these operations have been concluded and the play button is pressed, the tape is pulled out of the cassette and rethreaded. The driving torque in each of these modes is applied directly to the cassette reels.

The machine is completely automatic in its mechanical functions. Relays and solenoids are used, and associated with these and their electronic driving circuits are several interlocking features.

Two motors are used, a head and capstan motor and a threading motor. Several belts are used to connect the motors to the appropriate drive points, but there is one difference in cassette machines that must be observed. Several of these belts use the shiny surface instead of the dull surface as the operating side. It is important to notice which way a belt is installed if service or maintenance is contemplated.

Type 2 Transport

The Type 2 mechanism incorporates an improved tape transport and threading ring. It is used in the higher-quality industrial and broadcast cassette machines. It has a toothed threading wheel that

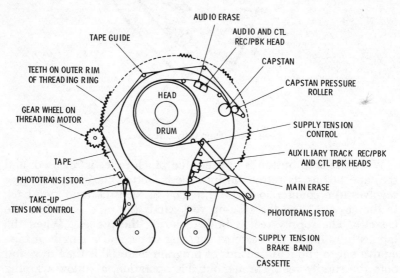

Fig. 14-5. Parts of Type 2 mechanism.

is directly driven by the dc threading motor. Fig. 14-5 shows the threading mechanism and the slightly different tape path used.

As the tape emerges from the cassette supply reel, it is pulled by the supply tension regulator arm to contact the main erase head and the auxiliary head stack. In the rewind and fast-forward modes,

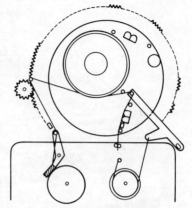

Fig. 14-6. Tape path for fast
forward and rewind.

the tape is not fully retracted into the cassette, but is left partially around the threading path. A separate motor is used to drive the tape spools in the shuttling modes and to keep a definite amount of tension in the tape. This is set so that contact is still made with the main erase head and the auxiliary heads (Fig. 14-6) to allow the control pulses and the time code to be read from the auxiliary track and fed to a remote editor when one is used.

Further refinements in the mechanics of these machines are:

1. Brushless dc motors are used to permit tight servo control. Typical raw time-base errors are around ¼ horizontal line.
2. Solenoids are used for engaging all the main operational and mechanical functions.
3. Remote operation is possible, and the necessary control lines appear at a special multipin plug on the rear panel.
4. Improved chroma circuits, with comb filters instead of ordinary passive filters, are used.

Fig. 14-7 shows the Sony VO 2860 editing machine, which has been used extensively in teleproduction and broadcasting, and Fig. 14-8 shows the BVU-200, which was developed specially for broadcast use. Several portable versions have been produced. These use 20-minute cassettes, and they can be run from internal batteries, or connection can be made to a separate power pack. The power pack may be used to run the machine from the 120-volt ac line and to charge the batteries. The machine is light enough to be shoulder carried along with a small portable color camera for applications such as electronic news gathering (ENG). These machines can record a full-color signal, but they play back only a low-resolution monochrome picture into the camera viewfinder or a separate monitor. To see color, an extra color pack is required. Fig. 14-9 is a photograph of a deck, color pack, and power unit.

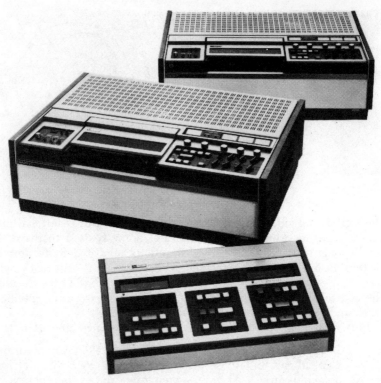

Fig. 14-7. U-matic editing vtr.

HEAD DRUM

The head drum is in two parts, a lower stationary section and an upper rotating section. Two heads are mounted on the periphery of the upper section and not on a bar as in many machines. The tape wrap is such that the heads are in contact with the tape in excess of 180°, thus giving a small overlap time in which to effect head switching. The signals from the heads are coupled to the playback preamplifiers by rotary transformers. The primary rotates with the drum, and the secondary is fixed; coupling is accomplished by a ferrite enclosure with a very small flux gap. Two head-tach coils are used for head switching, and one is also used for servo feedback.

The drum is belt driven from an ac motor that also belt drives the capstan at a constant speed. The head-drum rotation is controlled by a magnetic brake.

The tape is ¾ inch wide. The arrangement of the tracks is shown in Fig. 14-10.

Fig. 14-8. U-matic vtr for broadcast use.

Fig. 14-9. U-matic portable vtr (center), color pack (right),
and power pack (left).

ELECTRONICS

The electronic portions of cassette machines are similar to those
of other vtr's; the main difference is the inclusion of control circuits
to effect the tape threading, which is now fully automatic. The circuits can be split into seven distinct sections:

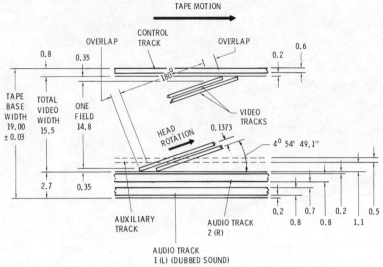

Fig. 14-10. Recorded tracks on cassette tape.

1. Video recording
2. Video playback
3. Color record and playback circuits
4. Servos
5. System control
6. Tv tuner and rf output
7. Audio record and playback

Video Recording

In common with all other vtr's, an fm system is used. It employs a square-wave multivibrator set to swing from 3.8 MHz for sync tips and no signal to 5.4 MHz for peak white (Fig. 14-11).

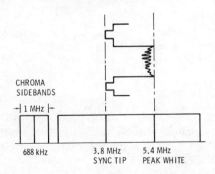

Fig. 14-11. Frequencies recorded on cassette tape.

The recording process is straightforward and simple, with the output of the modulator transformer coupled to the head driver amplifiers. A separate amplifier is used for each head, and a control is provided at the input of each amplifier to optimize the head current. The output of the modulator is balanced for symmetry and passed through an afc circuit that produces a dc output to offset any change in the sync-tip frequency.

At the input stage, a sync separator is used to provide vertical sync for the servo reference, a delayed burst gate for the automatic color control, and a reference pulse for the luminance agc system. The color is separated and treated separately, as described in a later section.

Video Playback

Many machines are made for playback only, but their playback circuitry is the same as that in the recording machines. Fig. 14-12 is a block diagram of the playback preamplifier section. The outputs of the heads are coupled by rotary transformers to individual pre-

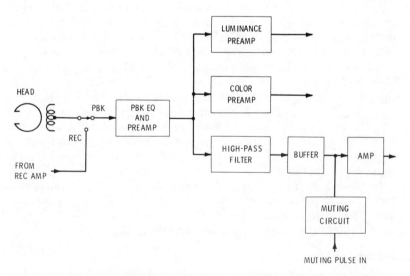

Fig. 14-12. Block diagram of playback preamplifier.

amplifiers. These amplify and equalize the signals to provide a flat response throughout the rf spectrum. The low-frequency color signal is taken from this point and routed to the color circuits, and the luminance fm signal is passed through a high-pass filter to remove all the color information. The fm signal is then amplified, and the output of each amplifier is alternately switched to the input of the limiters. The switching is accomplished electronically and is controlled by pulses from the two head-tach coils. It is timed to occur 5 horizontal lines prior to the vertical-sync interval.

The limiter is contained in one IC, and its output is differentiated into spikes as in other vtr's. The highest frequencies are transformed into the most negative part of the signal; hence white appears as the most negative part in the output. This signal is inverted, the processed color is added, and the result is routed to three places. The first is a sync separator, which is used for burst keying on playback; the second is the rf modulator; and the third is the video output jacks on the machine.

A muting circuit is added either at the preamplifiers or in the video output stages. This is controlled by an audio signal that can be recorded onto the audio tracks, and it allows audio and pictures to be synchronized for display purposes and demonstrations.

Color Record and Playback Circuits

In the record mode (Fig. 14-13), the 3.58-MHz color signal is separated from the luminance signal at the input to the machine and passed through a bandpass filter to reject everything outside the chroma band of 3.58 MHz ±500 kHz. This signal is mixed in a balanced modulator with a stable 4.27-MHz signal from a Colpitts oscillator. The difference frequency of 688 kHz is separated by a filter and applied to the heads in parallel with the luminance fm signal. The fm signal acts as a bias signal for the 688-kHz signal. A low-pass filter limits the top of the color signal range to 1.3 MHz, and the color signal level is maintained constant by a fast-acting agc. This ensures that the color signal is kept out of the spectrum required by the fm signal, and that a correct ratio between the two signals is maintained.

In playback (Fig. 14-14), the 688-kHz signal is separated from the fm and processed separately to remove the time-base errors and jitter components. The method of achieving this is covered more fully in the chapter on color recording.

Servos

The basic cassette machine is not designed to play back into a television system in synchronization with other equipment, and neither is it expected to perform editing. Consequently, the servo employed is of relatively simple design.

The capstan is driven by a constant-speed ac motor, which is also used to belt-drive the rotating head drum. In record, the incoming vertical sync is used as the reference signal, and the head-tach pulses are the feedback signal for the simple servo that controls the magnetic brake on the head drum (Fig. 14-15). The vertical sync is also

294

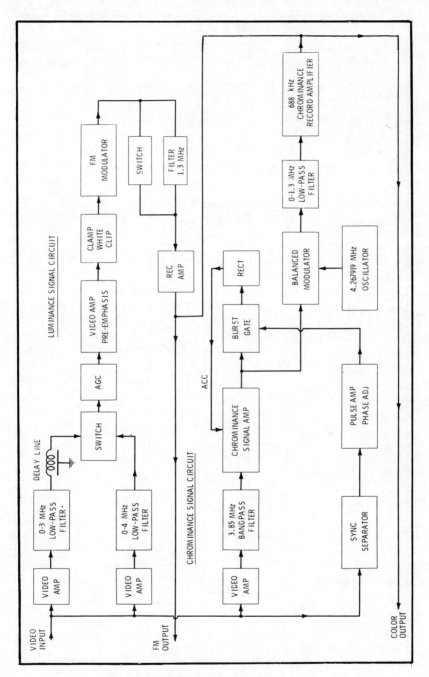

Fig. 14-13. Block diagram of record mode.

295

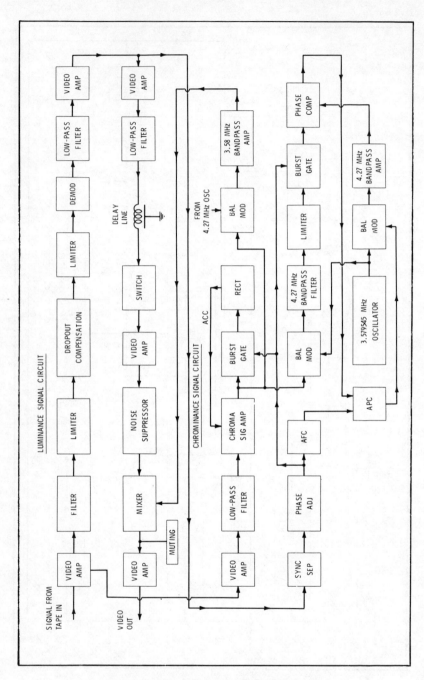

Fig. 14-14. Block diagram of playback mode.

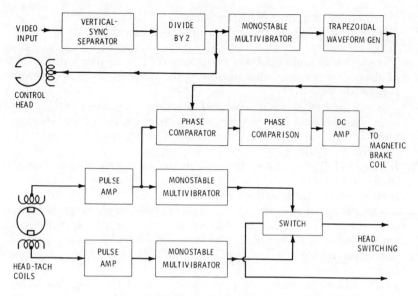

Fig. 14-15. Block diagram of servo.

used to form the control-track pulses, which are recorded at a 30-Hz rate.

In playback, the servo is a little more complicated. The control track is used as the reference, and the head-tach pulses are again the feedback information.

The servo circuit is also used to provide the head switching, manual tracking control, blanking pulses for the dropout compensator, and detection of the control-track pulses or signals used for video muting.

System Control

All of the mechanical functions, such as tape threading, tape transport, auto stop, etc., are governed by a system of control circuits. Nine distinct modes of operation can be initiated after the power has been turned on, and these are listed in the service manuals as:

1. Play
2. Manual stop during play mode
3. Auto rewind at tape run-out during play
4. Rewind
5. Manual stop during rewind
6. Auto rewind at tape run-out during play or fast forward
7. Fast forward

8. Manual stop during fast forward

9. Auto stop at tape run-out during fast forward

The service manual contains a separate functional diagram for each of these, with a brief description of the mechanical and electronic actions they control.

Every one of these mechanical functions is started and stopped by a solenoid that moves a mechanical part, or by a relay that provides or interrupts power to a circuit. These devices are driven by transistors, diode gates, other relay and switch contacts, and the front-panel buttons. All the circuits include lock-out and safety features to inhibit all nonselected functions and thus protect the tape when it is out of the cassette.

For the purposes of illustration, the play mode only is shown (Fig. 14-16), but it is typical of all the other modes. In this mode, several items are controlled. The tape threading is started, the head and capstan motor is powered, the pinch roller is closed, tape motion is started, back tension is applied to the tape, the cassette is firmly locked in place, all the buttons except the stop button are mechanically locked out, and in the final threading phase the stop button is electronically disabled.

TV Tuner and RF Output

Machines with a record capability are most easily recognized by their slightly larger size due to the addition of the record circuits and a tv tuner on the right-hand side of the machine.

The tv tuner is a standard vhf-uhf tuner of the type found in any domestic tv set. Normal vhf and uhf antenna terminals are provided at the rear of the machine. Channels are selected in the normal manner, and fine tuning is accomplished by a fine-tuning ring and a simple meter labeled as a tuning indicator. The output of the tuner passes through a set of relay contacts and directly to the output of the machine, which is connected to the tv set or monitor used to view the program. When the machine is powered, the relay pulls in and connects the tuner to the input of the record circuits and the output of the machine to the monitor or tv set. In this manner, it is possible to record and immediately play back an off-the-air signal in color.

An rf modulator is placed at the output of the machine, and it connects to the standard antenna terminals of a tv set. The input to the modulator contains the video signal from the machine together with the left, right, or mixed L + R audio channels. The output is a standard tv signal on a channel that is vacant in the area in which the machine is to be used. Generally, this can be any of the channels 2 through 6. Usually, two channels are provided, and the

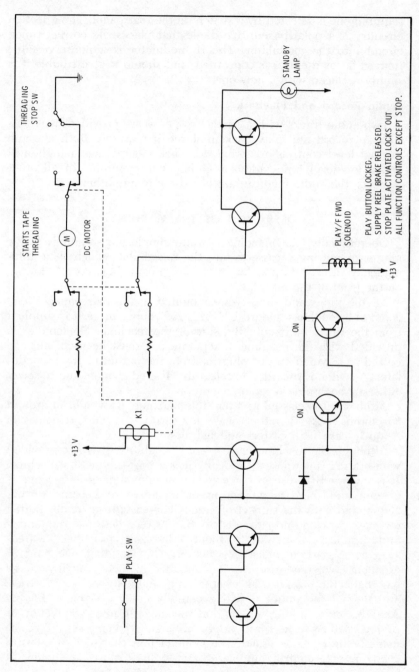

Fig. 14-16. Simplified circuit of system control.

299

required one is selected by a switch on the rear panel. The selection circuitry is a polarity-sensitive diode that places the correct tuned circuit into the modulator. The rf modulator is a nonserviceable item as far as the user is concerned, and should it give trouble it is simply replaced with a new one.

Audio Record and Playback

Two audio tracks with a 12-kHz response are provided; it is possible to record and play back in stereo if required. Both agc and manual level controls are provided. Each channel has a high-level and a low-level input and a high-impedance output. In all other respects, the audio facilities are like those in any other vtr.

OPERATION OF THE MACHINE

Operationally, the machine is quite simple. Operation is just a matter of putting a cassette into the raised slot, which then automatically drops into position, and then pressing the correct button on the front of the machine.

All the plugs and sockets are mounted in the rear panel. Video is fed in through a standard UHF PL 259 connector, or it is supplied from the tuner built onto the side of the machine. No monitor is provided with the machine; a separate one must be used and can be fed from video output plugs or from the output of the rf modulator. A switch is provided to select the desired channel and to select either monochrome or color operation.

Although it is classed as a portable machine, its weight of around 65 pounds makes it rather heavy for carrying around. It comes in a sturdy case with casters and handles.

Although the machine has a record capability, it is not provided with editing functions, as basically it is a post-production machine. It is intended for use in places where a minimum of knowledge is needed and there is no reason or incentive to become further acquainted with the subject of video. The cassette is ideally suited for mass copying and mass distribution of recorded material, and it finds application in entertainment, education, and specific areas such as sales or technical instruction. The fact that one hour of excellent color recording can be contained in such a small package has made the cassette an acceptable medium to many industries, and there is no doubt it will eventually enter into many homes. Already, multicopying centers that specialize in mass production of prerecorded cassettes for a variety of clients are being set up. Every manufacturer has accessories or modifications that will permit still frame, remote control, automatic rewind, back space, and pause to be included.

THE SMALLER CASSETTE MACHINES

There are a few varieties of cassette vtr's that are much smaller than the U-matic machines and that were originally introduced as consumer items. Typically, these use ½-inch tape in a completely different format from the U-matic or the open-reel machines. One of these is the Betamax format. It is made by several manufacturers and will be covered briefly because it contains some very unusual ideas.

Fig. 14-17 shows an example of a Betamax machine and its cassette. Fig. 14-18 shows the tape path used. Note that it is the mirror image of the U-matic tape path. The cassette is similar to the U-matic cassettes, but it measures only 6 in × 3½ in × 1 in, and the machine in Fig. 14-17 is about 24 in × 7 in × 16 in and weighs just under 41 pounds. A tv tuner and an rf modulator are included to allow off-the-air recording, playback into the antenna terminals of a tv set, or recording of one program while another is being watched. (Although off-the-air recording is possible, all manufacturers warn that it may contravene federal copyright laws.)

The single-speed models have a tape speed of 4 cm/s and will give one hour of recording. The tape speed of the dual-speed models can be halved to 2 cm/s to give two hours of recording.

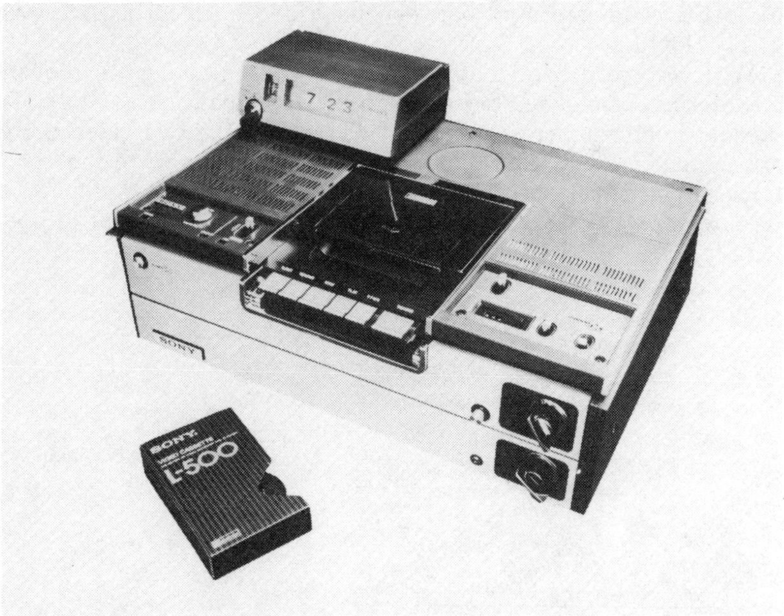

Fig. 14-17. Betamax vtr and cassette.

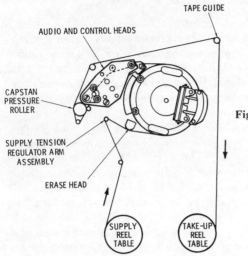

TAPE GUIDE

AUDIO AND CONTROL HEADS

CAPSTAN
PRESSURE
ROLLER

Fig. 14-18. Betamax tape path.

SUPPLY TENSION
REGULATOR ARM
ASSEMBLY

ERASE HEAD

SUPPLY
REEL
TABLE

TAKE-UP
REEL
TABLE

The slow speed and gentle tape handling in these machines allow a very thin tape to be used. This is one reason why the cassette is so small and why so much program material can be put onto one cassette. But the great tape-saving feature of this format is that no guard bands are used between the video tracks. This is shown in Fig. 14-19.

With no guard bands, other measures must be taken to prevent mistracking, luminance cross talk, and chrominance cross talk. To prevent luminance cross talk and mistracking, the two video heads are canted 7° away from the 90° azimuth with respect to the video tracks (Fig. 14-20). At the fm frequency used, this 14° difference is enough to make a track appear invisible to the head that did not

A	TAPE WIDTH	1/2 in (12.65 mm)
B	VIDEO TRACK PITCH	58.5 μm
C	VIDEO WIDTH	10.62 mm
D	CONTROL TRACK WIDTH	0.6 mm
E	AUDIO TRACK WIDTH	1.05 mm

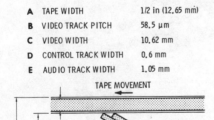

TAPE MOVEMENT

VIDEO HEAD MOVEMENT

TAPE VIEWED FROM COATED SURFACE

Fig. 14-19. Betamax one-hour track format.

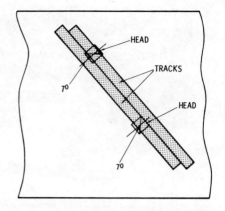

Fig. 14-20. Head orientation to reduce cross talk.

record it and thus effectively appear as a guard band between alternate tracks. On playback, the asymmetrical placing of the PG coils and the pulse vanes indentifies the heads and controls the servo so that each head plays back the correct track.

The chroma is heterodyned down to 688 kHz, as in the U-matic system. However, two differences are introduced in this process:

1. The down-converted frequency of 688 kHz is too low to be invisible to the head that is offset in azimuth, so an alternative method of avoiding chroma cross talk has been used. On record, the chroma phase is alternated every other horizontal line in the frame recorded by head A. The normal NTSC phasing is maintained when head B is recording. On playback, the cross talk from the wrong track has an adverse phase relationship to the chroma from the correct track. By delaying this signal for one horizontal line in a comb filter and comparing it with the subsequent line, the cross talk can be removed and the correct chroma signal strengthened without restricting its bandwidth.

2. A heterodyning frequency of 4.27 MHz is used to convert the 3.58 MHz to 688 kHz and back again. Unlike the case of the other machines, this 4.27 MHz is produced from a phase-locked loop, and so it is phase locked to the incoming horizontal-sync pulses in both record and playback. Thus the up- and down-converted chroma is also phase locked to horizontal sync.

Because of the reasonable quality of the picture from these machines and their low tape cost, several advanced models have been produced for use in industry and education. These have editing and other facilities added, along with an improved threading mechanism similar to the U-matic Type 2. Fig. 14-21 shows the

303

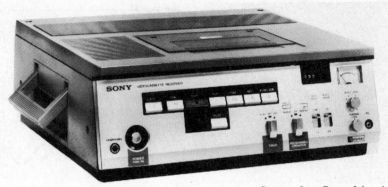

Fig. 14-21. Model SLO-320 Betamax vtr.

Model SLO-320, which has remote operation and all mechanical functions controlled by a simple memory circuit and seven solenoids.

Basically, the servos and control circuits are the same as in other vtr's and are quite straightforward, but slight differences do exist between the models. A few minor ways in which the Betamax machines differ from the U-matic machines are:

1. The tape automatically threads when the cassette is inserted.
2. All functions, such as rewind, fast forward, etc., are carried out with the tape run out of the cassette and around the threading path.
3. End-of-tape sensing uses a metallic foil that attaches the tape to the reels inside the cassette. The foil detunes an oscillator when it is pulled past a sensor just outside the cassette.

Chart 14-1 shows the main format specifications for the Betamax machines (two-hour format).

Chart 14-1. Betamax Two-Hour Format Specifications

Tape Width: 0.5 in
Video Track: 30 μm
Control Track: 0.6 mm
Audio Track: 1.05 mm
Head Gap: 6 μm
Tape Speed: 4 cm/s and 2 cm/s
Head Speed: 272 in/s
FM Frequencies: 3.5 to 4.8 MHz (approx)
Chroma Frequency: 688 kHz

15

Broadcast Helical VTRs

From the earliest days of vtr's, it was recognized that the helical format had certain advantages over the quad format. However, these advantages could not outweigh the seriousness of the time-base errors, tracking problems, and low-quality pictures of most helical formats. To obtain a broadcast-quality picture from a vtr, a high fm frequency must be used. Because there is a practical limit to the small size of the head gap, a high writing speed is required. Hence, high writing speed has always been synonymous with high quality. To obtain the high speed, short transverse tracks had to be used, and this also produced the other essential requirement of broadcast pictures—a stable sync signal on playback. As a result of all these factors, the quad became the broadcast machine and the helical did not.

As technology developed, the shortcomings of the early helical machines were overcome, and the helical format became practical for broadcasting. Before we discuss the broadcast helical vtr and how it came to be developed, a comparison between the quad and helical systems is desirable.

Broadcast VTR Requirements and Formats

The main advantages of the quad system are the high picture quality and the very low time-base errors. For a long time, these two factors made it the only vtr capable of broadcast performance. However, the quad vtr has several disadvantages that can be overcome only by careful adjustment of extensive and complex electronics. The heads tend to wear at different rates and have different frequency-response characteristics that cannot always be offset by the multiple preamplifiers. Geometic errors are always present to some degree, and so monochrome and color banding are never entirely absent. Because of the sharp angle at which the heads meet the tape, ferrite heads cannot be used. As a result, head wear is quite fast, and continuous checking and adjusting is necessary to overcome the visual errors mentioned above. This adversely affects both operating and production time. These errors naturally compound with successive copies (generations), and it is not long before picture degradation is quite noticeable. Also, the heads are expensive and must be changed about every 250 hours.

The quad machines are big and complex; they are costly to build, operate and maintain; and they use much tape. Operationally, it is impossible to obtain slow-motion or still-frame pictures, and so editing is difficult and time-consuming.

But despite all these disadvantages, quad vtr's can be made to produce excellent results. The quad machine was first required merely to record and play back a monochrome picture, and that is what is was designed to do. But as time progressed, more capabilities were required; color had to be added and color editing made possible. Production requirements seem to be pushing the quad machine to the limit of its operational abilities. Hence, an alternative is very desirable.

The only alternative to the quad vtr is the helical machine, which has several advantages over the quad in both the production and engineering sense. Production advantages are the ability to produce still-frame and slow-speed effects, and even a picture moving in reverse. These capabilities make editing easy and much less time-consuming than with quad machines. The engineering advantages are also important. Because it is possible to use a single head to record and play back an entire frame of video, the problems of picture banding and geometric errors are entirely absent. Only one playback preamplifier is needed; hence the setup is much easier, and many other electronic circuits are no longer necessary.

Because ferrite heads can be used, the heads have a much longer life, and picture deterioration due to head wear is much less than with the quad machine. Also, the heads are less expensive and need fewer operational adjustments. By using precision mounted heads, the penetration can be kept constant over a long period of time, thus improving the overall performance and minimizing the setup procedures.

Although guiding the tape around the heads is extremely critical, fixed ceramic guides with a long life can be used. Because of this, the head and guide adjustments and the air system that are necessary parts of the quad vtr are not required in the helical machine.

All of these advantages lead to a smaller overall deck that is cheaper to build and maintain and requires much less time in overall production operations. However, the basic helical vtr has several disadvantages that completely overshadow its advantages, and until recently these have not allowed it to produce broadcast-quality pictures.

To produce broadcast-quality pictures, a high writing speed is needed, and if the whole field is to be recorded in one track, that track must be very long. Even by reducing the head gap to the minimum attainable, a track length of 16 to 17 inches is still required. Pictures with sufficient frequency bandwidth can then be produced, but this long track produces the main problems that have beset the helical machine and are entirely responsible for its nonbroadcast performance. The three main problems are:

1. It is extremely difficult to guide the tape accurately around the head drum in such a long path, and hence the track is often not perfectly straight. This produces tracking errors and makes interchange difficult to achieve.
2. The slightest timing errors in the drum rotational speed and the tape travel have devastating effects on the final picture.
3. One of the biggest problems with the single-head helical vtr is the time when the head must leave the tape as it crosses from one edge to the other. This, along with incorrect tension, produces "skew" errors.

Despite all its advantages, the first two points alone prevented the helical vtr from becoming a broadcast machine because they were almost impossible to overcome. Various methods for correcting these faults were demonstrated, but the resulting machine was as expensive and complex as the quad machine and thus could not compete with the well-established standard.

The simple way to improve the timing errors and guiding problems is to reduce the track length. But then the writing speed must be reduced if one field is still to be placed in one track. Hence this method leads to a picture of less than broadcast quality in terms of frequency response and bandwidth, and it dictates that color-under methods must be used.

THE REQUIREMENTS FOR A BROADCAST HELICAL VTR

If the disadvantages of the helical machine could be overcome, it would be possible to incorporate several desirable features that cannot be added to the quad machine. For example, the quad format has been standardized with 10-mil video tracks, and the spacing of the tracks must accommodate this width. The original choice of the track width was governed by the manufacturing techniques at the time. If a 5-mil track could be used, a great saving in tape could be achieved by making the tracks closer together. Also, a quad machine cannot use a narrow tape. If it does, its track must be shortened, and then its head speed must be decreased, resulting in a loss of quality.

In a helical machine, because of the small angle the tracks make with the tape, a narrower tape can be used. Also, narrower tracks with close spacing can be used. Both together would produce considerable savings in the amount of tape used. Narrower tape costs less, takes up less library space, and is much cheaper to ship; hence, the overall tape cost could be reduced greatly. A lighter, narrow tape would lead to a simpler mechanical deck, thus reducing the cost of the machine. These arguments, added to a threefold improvement in head operating cost due to the longer life, make the helical vtr a highly desirable device.

Consequently, work on the helical machine never ceased, and finally it reached a stage of development at which it was capable of broadcast performance. At this point, the desirable features of such a machine were reviewed, and the following list contains the most important of these.

The video performance must be at least as good as that of the quad machine, and preferably better.

The entire signal must be recorded and played back without any loss of parts or insertion of irregularities.

The active or visible part of each field should be recorded and played back by one head only.

The time-base errors and skew must be very low.

High-band fm and direct color must be used.

There must be no banding or geometric errors in the picture.

Interchange must be excellent.

Multiple generations must be possible.

Head life must be as long as possible.

Operation and maintenance must be easier than with the quad machines.

'Still frame and slow speed in forward and reverse must be possible.

Editing must be very easy.

At least two high-quality audio tracks and a third separate track for cues and time codes should be available.

A protected control track must be provided.

Head change must be easy and quick.

Tape cost must be less than for the quad format.

Purchase price must be less than for a quad machine.

The machine must be smaller than a quad machine.

A portable version should be possible.

Several helical machines of broadcast quality soon appeared, all of which met most of the listed requirements.

THE HYBRID, OR SEGMENTED, HELICAL VTR

By looking closely at the faults and advantages of the quad and helical machines, it is possible to choose a combination of both and produce a machine that seems to provide the best of both formats. The result of this combination is the hybrid, or segmented, helical video tape recorder.

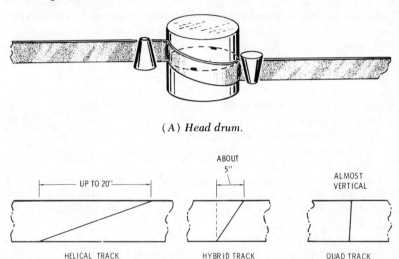

(A) *Head drum.*

(B) *Comparison of tracks.*

Fig. 15-1. Hybrid head and tape layout.

The basic hybrid format is a helical machine with several quad characteristics. Fig. 15-1 shows the tape wrap and a comparison of the tracks on the tape for the three formats.

In the hybrid machine, the tape is wrapped in a helical fashion slightly more than halfway around a head drum of very small diameter, to give very short slanted tracks. The tracks are scanned by two heads mounted in a small disc. With two heads rather than four, there will be fewer geometric errors, less electronics, and less banding in the picture; and these faults will be easier to correct. By segmenting the field and including about 50 lines in each track, a very high scanning speed can be used. This allows a high-band signal to be used for the reproduction of broadcast-quality pictures with direct color recording.

Using a narrow tape results in a small tape-to-metal contact area. This produces low friction around the drum and thus less tension difference in the tape, with the result that skew and raw time-base errors are very low (about 1 to 5 microseconds).

The low-mass, low-inertia head disc tends to run at a uniform speed and is easy to servo control, so it is not a major source of time-base errors. In fact the gyro error with this head disc is so low that this format is ideal for mobile work, and early models found extensive use in military aircraft.

Because the heads still meet the tape obliquely in helical fashion, ferrite heads can be used and provide all their attendant advantages. Precision fixed guides can be used, and because of the very short path around the head drum, the interchange capability is excellent. This also makes a vacuum system like that of the quad machine unnecessary.

The overall result is a small machine that is easier and less expensive to build, easier to operate, and capable of having more operational facilities than the quad machine. Its main drawback is that it loses the principal advantage of the true helical machine, which is the ability to produce slow-speed and still-frame pictures. To accomplish this, an auxiliary frame-store device is necessary, and this is a large and expensive addition.

The important point to be realized about the segmented helical vtr's is that they were all developed before the appearance of the digital time-base corrector, and they were all good enough to be considered a serious alternative to the quad vtr.

The three major machines designed as segmented helicals are described briefly in the remainder of this section. It is significant that two of them were designed mainly for the European tv systems. European tv is quite unlike tv in the United States in many respects. At one time there were four separate monochrome systems and two color systems. But now that the British 405-line system and the French-Belgian 819-line system have been phased out, Europe is left with the West European countries on a 625-line, 50-field/second PAL color system and the eastern countries on a 625-line, 50-field/second SECAM color system (a few exceptions exist). The quad machine has some problems that do not allow it to become fully integrated in either of these systems, and an alternative has always been desirable. Also, the economics of television in most countries is radically different from the American commercial system; in addition, the economics of major tv shows require world-wide distribution for survival. So, Europe with its large population is a very attractive market for the American show, and the U.S. market is very attractive to the European producer. These and other factors combine to make Europe an ideal market place for a new type of vtr.

The three major segmented helical machines are:

1. The Echo machine
2. The IVC 9000
3. The Bosch-Fernseh BCN machines

Each will be covered briefly.

The Echo Machine

The Echo machine was the first of the segmented helicals to be produced as a commercial machine. It was not intended to replace the quad machine, but to be used in areas where the more expensive quad was unnecessary. For example, a show could be edited on one or more of these machines and then the final tape transferred to the quad format for broadcasting.

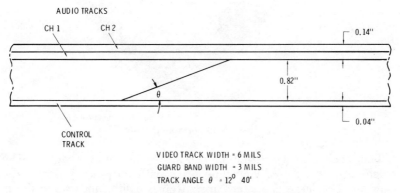

Fig. 15-2. Track layout in Echo machine.

In the Echo machine, the tape is wrapped 190° around the 2-inch–diameter head drum, and the two heads rotate in a horizontal plane to scan the tape. The drum assembly is manufactured with the heads and tape guides as integral parts. The total tape path from entry to exit is 6.7 inches, of which only 3.8 inches are in contact with the drum and only 3.8 inches are used for the track. Fig. 15-2 shows the track layout, including the positions of the audio and control tracks. The extra 10° of wrap allows head switching to take place on the front porch of the horizontal-sync pulse and allows a slight overlap at the switch point.

With a monochrome 60 Hz tv signal, the heads rotate to give a scanning speed of 1470 in/s, and six picture segments of 40 lines each are used. With a color signal, the head speed is changed to give seven segments. This change of speed and use of a different number of segments is a characteristic of the hybrid machines. It allows them to be easily adapted to any tv system by changing only a few electronic boards and without any mechanical changes.

Fig. 15-3. Echo portable machine.

The mechanical construction also has some unique features. The machine uses a closed-loop tape drive with no pinch roller at the capstan. The large-diameter capstan gives a firm and stable drive, and so the tape transport is very simple and free from the tape-path distortions that can be caused by an imperfect roller.

All the motors are printed-circuit dc types, which have low armature inertia and hence can be easily and accurately servo controlled. The small portable version shown in Fig. 15-3 uses only one motor with belt drives for the other functions.

In recording, the Echo machine uses the standard broadcast highband fm (7.06 MHz sync tips to 10.0 MHz peak white). Direct color recording is used.

The playback method is basically the same as that in the quad machine, which is basically a burst-track method. The time-base errors are corrected by a horizontal locked loop, so the head is tightly controlled. A control tone of 1.5 times the burst frequency is recorded on one track and is used in playback to correct velocity errors. Both magnetic and photosensitive tachometers have been used for head control.

Operationally, the machine will both assemble and insert edit in color with a complete frame match in all modes. It uses 1-inch tape with 15 in/s as the standard speed, but others can be used. A program time of 1½ hours is possible with a 12½-inch reel. An electronic tape timer measures hours, minutes, and seconds, and indexing is repeatable to 0.1 percent. Hot-pressed ferrite heads with a rated life of 1000 hours are used for the video.

The heads are changed by changing the entire head-drum assembly. All the assemblies are made the same and are completely interchangeable. This was done to make head changing in an aircraft or in other difficult situations easy.

Although originally designed for studio use, this machine found its first use in military aircraft and was used in this application for a number of years, to the almost complete exclusion of other uses. Later, the machine was introduced to the broadcast industry in a deck suitable for studio use.

The IVC 9000 Series Machines

Although an advocate of the one-head, full-wrap system for many years, IVC made a departure from this philosophy to produce a segmented helical machine. It was developed in association with Rank Industries in Britain and Thomson in France and was intended primarily as an alternative to the quad machine in the European tv systems.

This vtr has two heads and a scanning speed of 1500 in/s. The tape is two inches wide, and with a linear speed of 8 in/s a 4800-foot reel gives 2 hours of recording time. The machine measures 49 × 28 × 62 inches, so it is comparable in size to the quad machines. The reels are mounted at waist height for easy loading, and all the monitoring points are at eye level (Fig. 15-4).

The head drum is smaller than in the other IVC helical machines, with only a 5¼-inch wrap of tape. The picture is divided into 54-line segments, so there are 5 segments per frame. The track layout is shown in Fig. 15-5. The cue track is an audio track of lesser quality than the main audio tracks and is intended for cues and codes. The control track is located inboard from the edge of the tape, and a track near the edge is provided for the SMPTE code. All of these tracks can be erased and recorded separately and do not interfere with the video. A full range of editing capabilities is provided in this machine.

The motor mounting board is made of a "waffle" construction to give added rigidity to the transport. The tape tension is kept constant by vacuum columns for tape loops and servo-controlled reel

Courtesy International Video Corp.

Fig. 15-4. IVC-9000 machine.

motors, with optical sensors on the tape loops. This system maintains a good tension control in the fast-forward and rewind modes, which is one of the main problems with the quad machines and some helical machines. The transport system allows the use of a nonmetallic capstan, which can drive the tape without a pressure roller. This produces an even drive and removes one extra place of tape contact.

Hot press ferrite heads with a rated 1500-hr life span are used for the video and erase functions. These have a frequency response up to 18 MHz, which permits fm frequencies of 9 to 12 MHz to be used. This is ideal for the PAL and SECAM color systems, which use higher frequencies than the NTSC system.

314

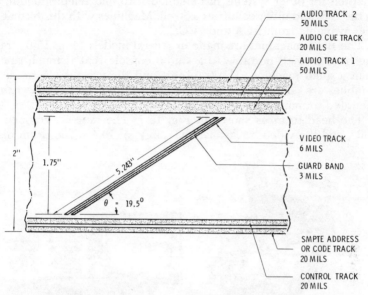

Fig. 15-5. Track layout in IVC-9000 machine.

The time-base errors are corrected electronically to ±1.6 ns by internal delay lines. The machine can be advanced or retarded by 64 microseconds (about one horizontal line) in steps of 127 ns. This is controlled by a 7.875-MHz master oscillator that also acts as the reference for all servos. This frequency was chosen because it is an integral multiple of the horizontal frequencies in both the 525- and 625-line systems.

To prevent banding in the playback mode, the head outputs are compared during the burst time and are set to be equal. After head switching, a chrominance-to-luminance ratio is set, either manually or automatically.

A total of four servos are used. These are the head-drum servo, capstan servo, tension servo, and reel servo. All are very tightly controlled and are much more complicated than those found in the simpler helical vtr's. The head drum is driven by a dc motor that has a high-resolution optical disc attached to it for servo feedback control. This disc has 40 lines scored in it, and at the proper rate of rotation it produces a 6-kHz signal. The capstan motor is a permanent-magnet dc motor; it has a similar disc with 1500 lines.

The Bosch-Fernseh BCN Machines

The BCN system was developed in Germany and was originally introduced into the European tv scene. The fact that it can be easily

315

adapted for other systems has enabled it to find teleproduction use in the U.S. and other countries as well. Machines with this format are also available from RCA and IVC.

The BCN machines are made in several models (Fig. 15-6), ranging from a small portable to a studio console that is much smaller than a quad machine. The transport and the various electronic assemblies are contained in separate chassis, and the machine can be mounted in many configurations.

The head drum is shown in Fig. 15-7. The tape is wrapped for 190°, and the video tracks make an angle of 14.4°. The drum diam-

Courtesy Robert Bosch Corp.

Fig. 15-6. Examples of BCN machine.

Fig. 15-7. Head drum in BCN machine.

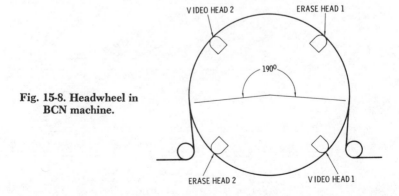

Fig. 15-8. Headwheel in
BCN machine.

eter is 50.3 mm (1.98 in), which produces a track length of 80 mm (3.15 in). The rotating speed of the head is 150 revolutions/second (9000 r/min), which gives a writing speed of 24 m/s (944.88 in/s). Each track contains 52 lines.

Fig. 15-8 shows the headwheel that spins inside the drum. The two video heads and the two flying erase heads are all mounted at right angles to each other. A rotary transformer, mounted on the shaft below the disc, is used for each head. This headwheel is easily removed and replaced, and the estimated head life is between 500 and 1000 hours. The head preamplifiers and the head motor are mounted in the head drum.

The entire head assembly can be removed by taking out two bolts that hold it onto the precision mounting. A new assembly is simply pushed into place and bolted down. A connector makes all the electrical connections, and no soldering is needed. In fact, replacing the complete drum assembly is quicker than replacing the headwheel.

The two precision mounted guides on the drum control the tape entrance and exit, and head or drum removal does not affect the tracking. The control-track head is mounted in the side of the head drum during manufacture, and so the phase relationship between the control pulses and the video tracks is permanently set.

The track format is shown in Fig. 15-9. The video tracks are 160 micrometers (6.4 mils) wide, and the guard band is 40 micrometers (1.6 mils) wide. Two high-quality audio tracks are used with Dolby A equalization. A third high-quality audio track is intended mainly

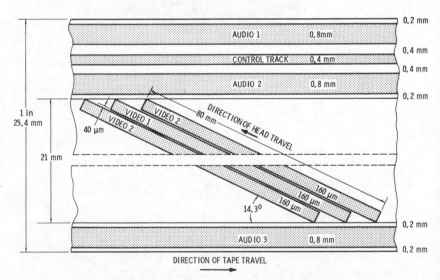

Fig. 15-9. Track format of BCN machine.

Courtesy Robert Bosch Corp.
Fig. 15-10. Two variations of BCN tape transport.

for cues, etc. The control track has been placed away from the edge of the tape.

The tape transport is shown in Fig. 15-10. The tape has a B wind (oxide out) and is guided by precision rollers at all times, The tape speed is 24 cm/s (9.45 in/s); tape motion is produced by a direct-drive dc motor with printed rotor. Sixty minutes of program can be obtained with 9-inch reels, but it is possible to fit 10½-inch reels. The tension in the tape is sensed by two arms that are attached to Hall-effect devices. These provide feedback to the reel servos to maintain constant tension at all times.

The raw time-base errors from the transport are on the order of 1 microsecond. A built-in analog TBC using switched delay lines and having a total range of about 9 microseconds reduces the errors to about 2.5 nanoseconds at the output. The circuitry includes a dropout compensator with a one-line delay and a velocity compensator with a one-line delay.

An interesting feature is that the timing-error voltage is displayed on the monitor. This appears as a series of short, nearly vertical lines laid over each of the video segments. The angle of the line is a measure of the raw time-base error. Two vertical lines, 9 microseconds apart, show the limits of correction and indicate if the servos are getting close to needing realignment or if a bad tape is being played.

Assembly and insert edits are possible with controls on the deck, or from a remote panel. In the insert mode, any combination of tracks can be selected. A useful feature is the ability to preview an edit. The edit can be set up and rehearsed, and the output of the machine makes the video switch so that the edit can be observed on a monitor but not recorded onto the tape. This allows checking and adjustment of the edit points before the edit is actually recorded. The LED timer readout used to select edit points is driven from a tape roller or from a built-in SMPTE time-code board. Both give frame-accurate counts.

The electronic provisions include circuits for chroma agc to eliminate the banding that occurs with any segmented system. Warning systems and indicator lights are included for the major inputs and for servo lock, etc.

Although still framing is impossible with this machine, a frame store can be added as an option. This device stores complete frames of video from the heads and changes the frames in quick succession; thus an exact frame can be identified for editing.

NONSEGMENTED BROADCAST HELICAL VTRs

The first nonsegmented helical machines to find serious broadcast use were the U-matic video cassette machines. These had to wait for the digital time-base corrector, and then they found use in areas where picture quality was secondary to picture content.

Electronic news gathering (ENG) is an example of this. The success of the U-matics was established when they were used by CBS News to cover President Nixon's visit to Moscow. An upgraded broadcast version was soon produced, and so was a high-quality portable. These opened the way for the larger open-reel helical vtr's to enter the field.

Although the digital TBC was the main development that allowed the nonsegmented helical vtr to appear as a broadcast machine, these new machines took full advantage of many other recent developments and appeared as full-fledged usable machines with many features built in rather than added on.

The Broadcast Helical Format

The format decided on by the SMPTE committee was worked out to use the advantages of two similar earlier formats, one from Ampex and the other from Sony. The resulting combined format is known as "Format C."

Ampex used their 1-inch helical format in a broadcast-quality machine called the VPR-1. It had several significant improvements over the earlier industrial and educational models. The Sony machine was new but was based on an earlier format. It used the "1.5-head" system. Although it was not compatible with the Ampex format, it was close enough to suggest that a combination of the two could produce an ideal helical format for broadcasting.

The SMPTE committee resolved the differences in the formats, and now both manufacturers produce machines to the same standard, with tapes from one completely interchangeable with tapes from the other. The remainder of this section will briefly introduce the original formats and then show the new format.

The Sony Format

The Sony approach was to use experience gained with a variety of earlier smaller machines. The most notable feature is the "1.5-head" system. This is a two-head system that had been used on some very early 1- and 2-inch machines. In many ways, this format is ideal for a helical machine that must give high performance.

A full wrap is used around the drum, and two heads scan the tape. One head is used to record and play back the visible part of the picture, and the other is used only for the vertical interval. About 4 lines of overlap are used at each switching point, but these are out of the visible part of the picture. Hence there are no dropouts, and a complete signal is recorded and played back at all times.

The drum circumference was about 16 inches, and this produced a writing speed close to 1000 in/s. This is fast enough to permit use of the broadcast high-band fm frequency range of 7.06 MHz to 10 MHz. The linear tape speed was about 9.63 in/s.

The main advantage of this format is that the visible picture is recorded by one head, and thus there is no banding or geometric distortion. Another advantage is that all special vertical-interval signals, such as VIRS and VITS, are recorded and played back by this same head. Thus any setups that use these signals will apply to all the active visible lines and not to just several segments of the picture.

To guide the tape around the very long path with a minimum of problems, an extremely high level of mechanical engineering is necessary. A very stable and rigid baseboard holds the head drum, guides, and motors. The head drum and the tape guide around it are precision milled by a computer-controlled numerical lathe to provide the required tracking and interchange capabilities.

The Ampex Format

One of the main changes Ampex made to their earlier designs was to use fixed guides at the drum entrance and exit instead of the movable types of the older machines. A very rigid baseboard was used to hold the motors and heads, and precision mounting of all parts was introduced. However, the tape wrap around the drum was not changed, and the single eccentric screw at the midpoint was still used to support the tape.

To overcome the tracking problems caused by the tape sag in this long path, an electronic rather than a mechanical solution was developed. It is called automatic scan tracking, or AST. A video playback head, separate from the video record head, is mounted on a piezoelectric device that is fed a sinusoidal voltage. This causes the head to oscillate from side to side as it scans the track. Through measurement of the rf level from the head, the edges of the track can be sensed, and this information can be used to servo control the position of the head as it scans the track. The head can be made to follow the most wildly misrecorded track and produce a satisfactory playback picture. The system also has the ability to produce slow motion and still frames, which are impossible to obtain on a long-track helical machine without this device.

The tracks on the tape were the same as in the older format, except that another audio channel was added. As before, only one head was used to record the video signal, and 10 lines were still dropped from the vertical interval as the head left the tape. These 10 lines were reinserted by the digital time-base corrector to give a complete signal.

The SMPTE 1-inch Type C Format

The SMPTE Type C format was devised as a practical compromise between the Ampex and Sony 1-inch formats. It retains the 1.5-head system and divides the tv field into two sections. The video beginning with line 16 in one vertical interval and ending with line 5 in the next vertical interval is recorded by one head. The vertical-interval track is recorded by the other head and contains the 10 lines not included in the video tracks. The tracks are placed on the tape as shown in Fig. 15-11. Note that the vertical-interval track is not in line with its associated video track, as was the case with the earlier machines using the 1.5-head system. This displacement leads to a reduction of the skew errors at the changeover points, which was one of the major design objectives for this type of vtr. Skew occurs only at the changeover points, and these are well out of the visible parts of the picture. An adequate track overlap is used so that there is no danger of lost information.

For those users who do not require the information contained in the vertical interval, the 10 lines of vertical-interval track and the head can be omitted, but nothing else is permitted in this place.

The rotating scanner has a maximum of six head locations:

 1 and 2. Video and sync record heads
 3 and 4. Video and sync flying erase heads
 5 and 6. Optional AST heads

If any of the optional heads are not used, dummy heads with the same protrusion are installed to minimize velocity errors and maintain the head drum balance.

Audio Tracks—Three audio tracks are provided, as shown in Fig. 15-11. Channels 1 and 2 are high-quality tracks similar to those of a

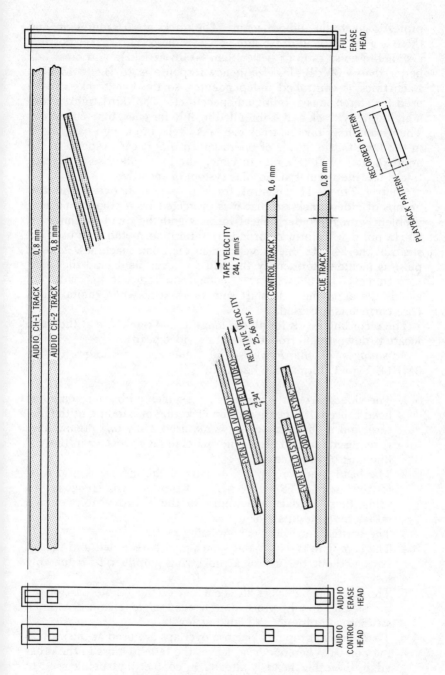

Fig. 15-11. SMPTE Type C tape format.

professional studio audio machine. The tracks are 0.8 mm wide, and there is a guard band of 0.8 mm between them. This makes possible a signal-to-noise ratio of better than 60 dB at 1 kHz and cross talk better than −60 dB. The frequency response is 15 Hz to 15 kHz. Each track is controlled independently, so the two tracks can be used for stereo music, bilingual speech, etc. The third track is also a high-quality track and is intended mainly for cues, time codes, etc. The preamplifier for this track can be switched to a high-band mode in the fast shuttle modes of operation so that it can respond to the approximately 100 kHz rate to which the time code rises. All three tracks use the normal audio bias system of recording.

Control Track—The control track is placed between the two groups of video tracks so that it is protected from edge damage—a problem with the older helical formats and the quad machines. A 30-Hz pulse waveform recorded by saturation methods is used instead of the 240-Hz sine wave used in the quad machines. An edit pulse is included to identify the odd and even fields and the alternate frames. This allows correct color frame edits to be made and avoids the annoying side shift sometimes seen when digital time-base correctors are used.

The recording heads for all the longitudinal tracks are at the same location, downstream from the rotating video heads.

Advantages of the Format—The following advantages of the SMPTE Type C format can be listed:

1. The video part of the signal is not segmented because only one head is used. Hence there is no flickering or banding of the picture, and no head matching is required. Only one preamplifier is required, and no electronic switching is necessary in the visible part of the picture.
2. The head used for the visible part of the picture is also used for the VIRS, VITS, CEEFAX, deaf captions, etc. Hence setups using these signals will apply to the entire visible picture rather than only part of it.
3. The vertical interval has no missing parts.
4. The complete tv video and sync signal is recorded and played back without the use of a processing amplifier to regenerate lost parts.
5. The head changeovers all occur out of the visible part of the picture. The only processing needed is to clamp the small switching transients to blanking level.
6. There are no dropouts because overlaps are used at the switching points. When one head leaves the tape to cross to the other edge, the other head is already in contact and being used to record or play back.

Other SMPTE Formats

The SMPTE has labeled the original Ampex 1-inch format the Type A format, and it calls the Bosch-Fernseh 1-inch format the Type B format. The European Broadcasting Union (EBU) has similar definitions for these formats when used in the 625-line, 50-field tv systems.

The Machines

The modern broadcast helical vtr delivers a level of performance that meets all the desired criteria. During the design stages, great attention was paid to increasing the video signal-to-noise ratio and decreasing the moire problems. This was achieved by using improved circuit design, ferrite heads, and high-energy tapes.

Typical specifications quoted for these machines are:

Video s/n ratio: 49 dB
Moire: −40 dB
Diff phase: 3°
Diff gain: 3%
K factor: 2%
Chrominance-to-luminance delay: 50 ns

Experience has shown that the quoted specifications are often conservative in practice, and that other important factors are also well within desired limits. In general, the interchange and tracking are excellent, the skew errors are minimal, and the raw time-base errors are extremely low.

Because of the high-energy tape and the low signal-to-noise ratio of the original recording, multiple-generation copies with a minimum of picture degradation are possible. Often, an original recording will have an s/n ratio better than 50 dB, and it is possible to make tenth generation copies that still have an s/n ratio of around 40 dB. Because most broadcast tapes are about fifth generation by the time they are aired, a low s/n ratio is of fundamental importance in teleproduction. Naturally, these machines must use a digital TBC, but even after ten generations the time-base errors are almost immeasurably small.

The slow-speed, still-frame, and reverse-direction modes are all controllable with knobs or buttons. Use of these functions makes editing much faster than the previous type of editing with the quad machine, with which a picture can be seen only at normal speeds. The counters and servos are accurate to within a frame, and edits are made on the correct color frame.

Because of the use of one video head, the setup and checkout for both recording and playback is simple and quick, and it can be performed with almost no preparation. All the servos and control circuits are either partially or entirely digital. Hence their checkout and adjustment is very easy, and they are not subject to drifting during operation.

In the fast shuttle modes, the equalization of the audio preamplifiers in the cue channel is switched out. The response normally rolls off at about 15 kHz, but when equalization is not used, the preamplifier has a wide bandwidth. This allows the time-code pulses, which are 2 kHz at normal tape speed, to be read at fast speeds, when their frequency rises to around 100 kHz.

Many standard digital ICs are used in the circuits, thus ensuring easy maintenance and minimum problems with availability of parts. All the circuits are mounted on PC boards that are easily accessible.

At the time of this writing, the two major manufacturers of machines produced to meet this format are Ampex and Sony. Other manufacturers are expected to provide models, but these will have many features in common with the two mentioned. These two will now be covered briefly, with emphasis on the important highlights of the machines rather than the features common to all vtr's.

THE AMPEX VPR-2

The model VPR-2 (Fig. 15-12) is the Ampex broadcast helical vtr. Originally, this was called the VPR-1 and was an advanced version of the earlier 1-inch helical machines, but with several new electronic and mechanical features added to make it capable of broadcast performance. Because of its close similarity of format to the original Sony BVH format, it was subsequently used along with the Sony machine as a basis for the SMPTE 1-inch Format C standard.

The VPR-2 has a full range of operational features required in a broadcast helical machine, including all edit functions, slow motion, and still frame. It can operate in a variety of positions and configurations, ranging from a small mobile unit to a full studio console with monitoring bridges.

Head Drum

The head drum is manufactured as a self-contained unit that can be removed from the front of the deck. It contains an integral dc motor, the head preamps and drivers, and the rotating heads. In the models manufactured before the Type C format was adopted, three heads were used: a record head, a flying erase head, and an AST playback head. These heads are mounted at 120° to each other in

Fig. 15-12. A VPR-2 video tape recorder.

the top part of the drum, and in operation the cross talk between them is about 45 dB below the peak video level. The record and playback heads can be optimized independently while the tape is running, thus allowing each to be set up for its individual function without shuttling the tape back and forth. A great operational advantage is that the tape can be replayed during the recording process to produce a full-quality video signal. The later models conform to the Type C format.

The head drum is provided in two versions, with and without the AST option. When the AST is not included, the AST playback head is replaced by a dummy head with the same protrusion to maintain the head-drum balance.

The tape is fed around the drum by fixed precision ceramic guides at the entrance and exit points, and a single eccentric screw at the midpoint of the tape path around the drum.

327

Tape Transport and Deck

The deck is made of a stable cast alloy board that has a ribbed construction to give it rigidity. This prevents mechanical distortions that can affect the tape path and ruin the interchange and tracking. The head drum, reel motors, fixed heads, etc., are all precision mounted onto the deck and are replaceable from the top or the front. The only exception is the capstan motor, which is removed from the rear.

The tape is guided around its path by precision rollers and fixed ceramic guides on subassemblies that mount onto the main deck (Fig. 15-13). The audio heads are manufactured with a fixed azimuth and require no adjustments once mounted.

A single capstan with a pressure roller is used to drive the tape around its path. This is a major change from the earlier Ampex models, in which there was no pressure roller and the tape passed

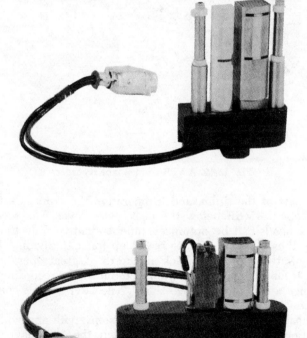

Courtesy Ampex Corp.

Fig. 15-13. Subassemblies with tape guides.

twice around a rubber capstan. An idler roller with a ribbed capstan drives the tape counter. The counter uses a wheel with many slits to chop a light source that shines on an optical sensor. The resulting pulses are counted and used to control the timers and other electronics.

Tape tension is sensed by a tension arm, which provides an input to the reel servos. Motion sensors for the reels work in all modes, and a differential brake is provided for each reel to protect the tape in case of power loss.

Automatic Scan Tracking

Automatic scan tracking, commonly known as AST, uses a special head assembly that can move in two planes simultaneously and thus allow a video head to be deflected from side to side as it scans a video track. During playback, the head can be controlled to follow any deviations in the actual track on the tape and thus eliminate all the tracking errors introduced by faulty servos, misaligned guides, or tape stretch. It can also be made to follow the slightly distorted track path caused by the slow-speed and still-frame modes. This technique serves to eliminate minor picture jittering, noise bars in the picture as a result of mistracking, and loss of interchange between machines.

The AST head is a normal playback head mounted on a piezoelectric device that is altered in shape and size by a voltage applied across its faces. (The highest voltage is about 160 volts.) The head can be deflected from the center of one track through the guard band to the center of either adjacent track and then back again within a single scan of one track, that is, during one rotation of the head drum. Since most tracking errors are much smaller than this, the head can easily follow all tracking errors likely to be encountered on a tape.

In the still-frame and slow-motion modes the track path is slightly different from the normal path followed by the head. The AST head can easily follow this new path during a single scan, and in the still-frame mode it actually jumps tracks at the correct point to provide a picture without the characteristic noise bar and without any vertical jitter.

In operation, a sinusoidal dither signal is fed to the piezoelectric device to produce a rapid side-to-side movement of the head. This produces a small amplitude modulation of the playback rf. The rf envelope is fed into a synchronous detector, which senses the magnitude and polarity of the error. The detector output is used as feedback information for the microposition servo, which controls the head. In the slow-motion mode, a field-rate ramp signal is also fed into the servo to control the head movement along the tracks.

329

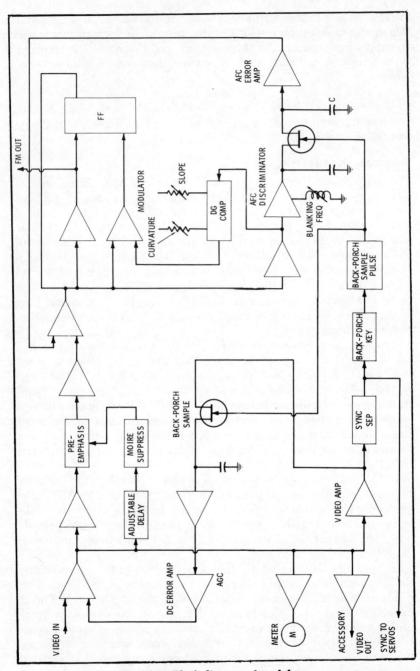

Fig. 15-14. Block diagram of modulator.

The AST uses a proper video head, and it produces no degradation in the video signal-to-noise ratio or in the time-base errors at the output of the machine.

Modulator Circuit

Fig. 15-14 shows a simplified block diagram of the video input and modulator circuit. The standard broadcast high-band fm is used with direct color recording.

Pre-emphasis and agc are applied to the input video signal as in other vtr's, and then the video is applied to the modulator. The fm signal is a square wave produced by two high-frequency op-amps and a flip-flop. A predistorter network is included in a feedback path to correct the differential gain errors introduced by this type of modulator. The afc circuit is a discriminator that produces dc from the fm. The output dc from the discriminator is sampled by a pulse developed from the sync stripper and the back-porch key circuit. The sampled dc is held in capacitor C, amplified by the afc error amplifier, and then used to adjust the gain of the video amplifier that feeds the modulator. The discriminator contains an adjustment to set the dc output level that corresponds to the blanking level of the video signal.

An interesting addition is the moire-suppression circuit. It adds a slightly delayed and corrected high-frequency signal back into the circuit at the pre-emphasis point.

Automatic gain control of the input is accomplished by sampling the back-porch level of a filtered input video signal. The dc resulting from this sampling is used to control the gain of the input amplifier. The video from this amplifier is fed to the meter circuit and to an accessory output as well as to the modulator circuits.

Demodulator Circuit

A simplified block diagram of the demodulator circuit is shown in Fig. 15-15. The input fm is converted to a differential signal by the input amplifier and then is passed through a chain of limiters to be demodulated by a low-pass filter. A feed from the rf input is used as an input for the dropout detector, which operates a switch in the output circuit to clamp the video to a preset dc level. Sync is stripped from the output video and fed to the other circuits in the machine.

Head Servo

The heads are driven by a dc motor that is an integral part of the drum. A simplified diagram of a motor-drive amplifier (MDA) is shown in Fig. 15-16, and a simplified block diagram of the servo that drives it is shown in Fig. 15-17.

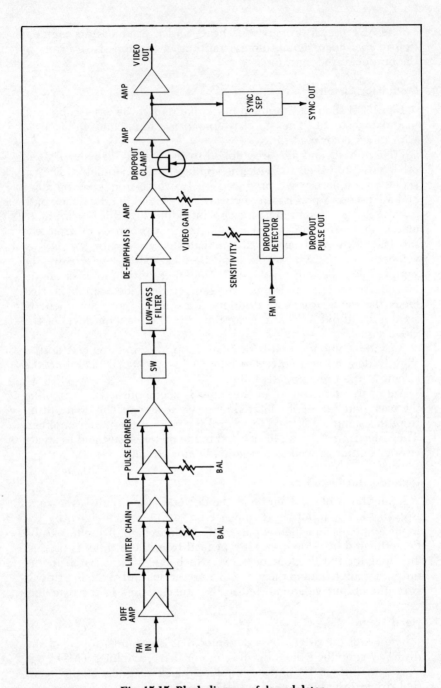

Fig. 15-15. Block diagram of demodulator.

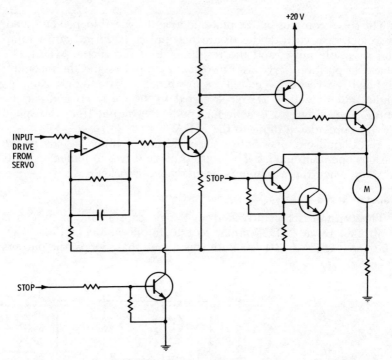

Fig. 15-16. Simplified diagram of head-drum motor-drive amplifier.

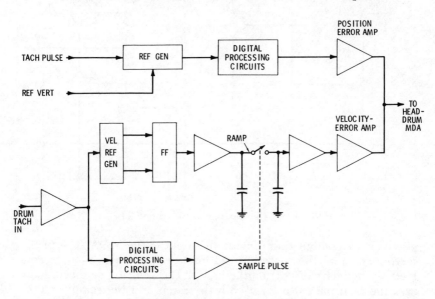

Fig. 15-17. Diagram of head servo.

The servo contains both a phase loop and a speed loop. The speed, or velocity, servo uses the drum tach pulses to form both a ramp and a sample pulse, and the resulting dc is fed to the MDA. The phase, or position, servo uses the head tach pulses and the incoming or playback vertical sync. These are processed through a complex digital circuit with a 2H clock signal to form a ramp and sample pulse. The dc output is combined with the output from the speed servo to provide an input to the MDA.

The digital circuits control the selection of input signals in the various operation modes. They also provide drives to indicator lights to show when the machine is properly servo-locked.

Capstan Servo

The capstan is belt driven from the dc capstan motor, which is controlled by an MDA similar to the simple circuit of Fig. 15-18. The servo (Fig. 15-19) has both speed and phase loops, the outputs

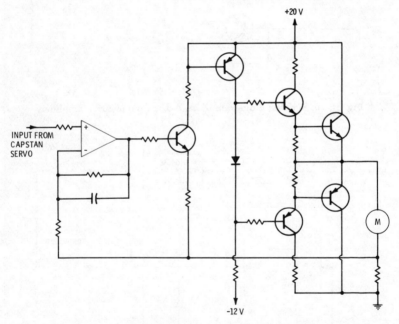

Fig. 15-18. Capstan motor-drive amplifier.

of which are combined to provide the input to the MDA. The phase servo loop forms the control pulses into a ramp, which is then sampled by the tach pulses from the motor. The velocity, or speed, loop uses the tach pulses to form both the ramp and the sample pulse. This loop also contains preset dc levels to provide a fast run-up to

334

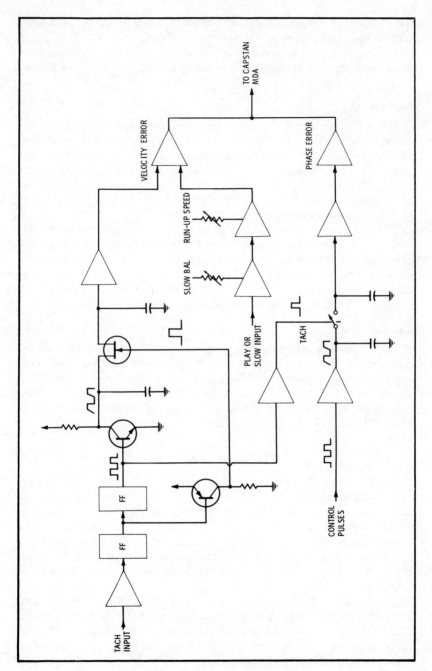

Fig. 15-19. Diagram of capstan servo.

335

speed when the tape drive is first engaged, and to control the capstan in the slow modes of operation.

The input signals are both digitally processed, and several control signals are used to route the signals and change the mode of operation.

Reel Servos

The reels are driven directly by servo-controlled dc motors. In common with the reel drives of other broadcast helical machines, these servos are driven and controlled by a complicated digital system with many inputs. A very basic circuit for one motor is shown in Fig. 15-20. However, each motor has different inputs in each mode of operation to control the slightly different functions each must perform in the various modes.

The main inputs are from the tape tach, tension control, and shuttle control. Other inputs are from the protection mechanism, the play and still indicators, etc. These signals are routed to each motor servo by a series of digital switches that are controlled by both the selected functions and the tape and mechanism sensors. A full description of the reel servos is impossible in a book such as this.

Control Panel

The control panel has a full range of operational controls and indicators that one would expect on a modern broadcast vtr. Full editing capability is provided. There are meters for the video and the audio. An LED tape counter shows hours, minutes, seconds, and frames for accurate editing and cue marking. Remote operation is also possible with the correct panel and connectors.

THE SONY MACHINE

The Sony broadcast helical machine is the BVH-1000 (Fig. 15-21). Fig. 15-22 shows the head drum, which is made in two parts, an upper and a lower assembly. The upper rotating assembly contains all the heads in its circumference. These are premounted during manufacture. The upper part has a diameter 50 micrometers larger than the lower static part. This allows a positive air passage between the tape and the large metal surface and prevents tape binding or "sticktion."

The lower part of the drum has a rabbet machined all the way around it to give exact and positive tape guiding for the complete wrap. This is an integral part of the drum and is made while the drum is being machined by a computer-controlled numerical lathe. This guide is raised 10 micrometers at its midpoint to compensate for the tendency of the tape to sag at the far point around the drum

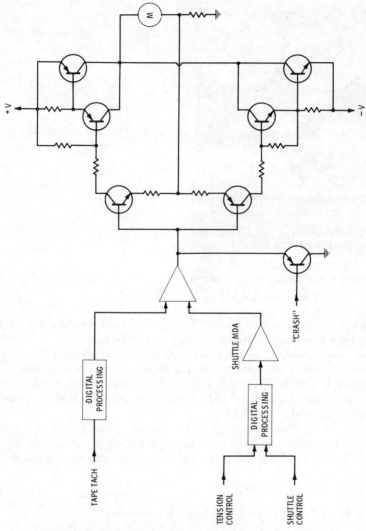

Fig. 15-20. Diagram of reel-drive circuit.

from the entrance and exit guides. It holds the tape variations to within 15 micrometers peak-to-peak to ensure the interchange of tapes between machines.

When the heads are replaced, the entire upper scanner is changed. The operation is simple, however, since only two bolts hold the scanner in place. Upon installation, the eccentricity of the drum and the dihedral of the heads must be checked, but special gauges are pro-

Fig. 15-21. BVH-1000 video
tape recorder.

vided to speed this procedure. The adjustment of record currents and playback frequency characteristics also can be accomplished readily. Only one head-tach coil is used, and this is mounted on the bottom of the assembly and not inside the head drum. When a new head scanner is installed, the PG signal phase must be checked.

The lower scanner is held in place by three precision mounted bolts. It is field replaceable.

Fig. 15-23 shows the tape path. The locations of all the guides were chosen to maintain optimum head-to-tape contact at all the fixed heads and keep the angle through which the tape turns to less

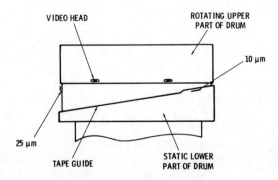

Fig. 15-22. Details of head drum.

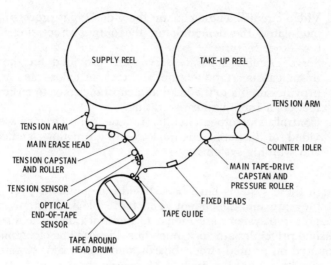

Fig. 15-23. Diagram of tape path.

than 40° at each guide post. The only ones that give a greater angle of turn are those at the head drum. All the guides and the capstans touch the coating side of the tape rather than the oxide; this is done to minimize tape wear and signal-to-noise degradation due to scratches or accidentally magnetized guides.

The most critical point with regard to edge damage and tape stress is the tape path to the scanner through the entrance guide. Here a slightly tilted scanner and an inclined guide are used to provide smooth tape flow and begin the small helical angle.

Five motors are used in the tape transport. There are two capstans, each with its own motor; each tape reel has its own motor; and there is a separate head motor. All use direct drives, and there are no belts. Three of the four servos in the machine are associated with the transport mechanism; these are the main capstan servo, the tension servo, and the reel servos.

In the record mode, the main capstan on the take-up side of the drum rotates at a fixed speed, and the tension capstan on the feed side is controlled by the tension servo. This keeps the correct tension in the tape at all times, especially around the drum. During the playback mode, the dual capstan system uses an auto skew control to minimize the skew error at the head changeover points. This dual capstan system is also used to help achieve fast field and frame locks when editing. The reel servos supplement the tape handling to provide smooth, even tape drive at all times. The servos are also used in the tape shuttling modes, as described later.

The main circuit systems in the BVH-1000 are:

1. Video circuits. These contain the video input processing, the modulator, the demodulator, the output processing, and a few other features.
2. Servo circuits. Separate servos are provided for the head drum, capstan, tape tension, and reels. A color-framing servo provides inputs to the head and capstan servos to ensure correct color framing during editing.
3. Controls and other circuits. Independent circuits are provided for the audio, cue tracks, editing functions, and control and warning systems.

Although these systems have been listed independently, to a large extent they are interdependent. The basic block diagrams of the video and servo systems are similar to those of quad machines, and established principles are very much in evidence. The electronics are all mounted on printed circuit boards, and each major circuit is on a separate board or combination of boards. Important test points and adjustment controls are mounted on the board edge for accessibility. Interconnections between the circuits are made on a PC motherboard, with harness connections to the rest of the machine.

Video Circuits

The basic video record system is quite straightforward. The video signal is clamped, pre-emphasized, clipped, and applied to a square-wave modulator. However, a few features are included that are worth noting. Fig. 15-24 is a simplified block diagram of the video modulator.

The burst is amplified to twice its normal amplitude before modulation. This is done to help correct time-base errors on playback and to get a better s/n ratio for the burst. Fig. 15-25 shows the basic circuit. A burst-gate pulse is produced from delayed horizontal sync and applied to the emitter of a transistor amplifier. This switches in the extra components to reduce the emitter impedance and thus increase the gain at this time.

The sync tips are clamped, but a keyed clamp is used. This ensures that any signals in the sync are ignored. One of the modern trends in tv is to insert digital signals into the sync pulses; digitized sound-in-sync is a good example of this. During the recording process, these signals must not be allowed to reach the modulator, and hence the keyed clamp is used.

The fm signal is produced by a line driver amplifier with feedback; Fig. 15-26 shows a simplified circuit. A slope detector in a feedback loop provides afc, which controls the sync-tip frequency of 7.06 MHz.

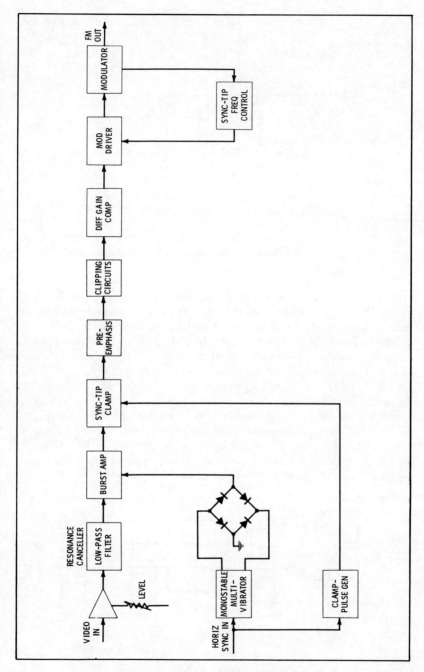

Fig. 15-24. Block diagram of BVH-1000 modulator.

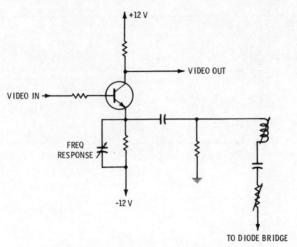

Fig. 15-25. Schematic diagram of burst amplifier.

This type of fm oscillator does not have ideal differential-gain characteristics, and to overcome this, the video signal is predistorted (Fig. 15-27). On playback, this produces a measured differential gain of about 2–3%, which is well within required limits.

The video-level meter is unusual. Two sample-and-hold circuits are used, one on the sync tips and the other on the horizontal back porch. The difference between these levels determines the sync amplitude, and this is used to provide the dc output of an op-amp to drive the video meter.

Demodulation is quite simple and is an example of how digital techniques are replacing the older analog methods in vtr's just as in other areas of tv. The fm signal is phase delayed by an RC network, and then the direct and delayed signals are applied to an Exclusive OR gate. The output consists of pulses of constant width but at effec-

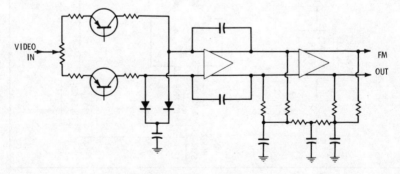

Fig. 15-26. Simplified modulator circuit.

tively double the fm frequency. These are filtered to give the video signal.

During the equalization process, a guard-band suppression circuit is switched in. This suppresses the noise bar that is normally seen in the picture during the fast shuttle modes. Thus an interference-free picture can be seen at all times during editing.

The Bidirex Dial

The Bidirex dial is a control for tape shuttling or fast forward and rewind. It has two separate modes of operation, called "shuttle" and "jog." Fig. 15-28 shows basically how this control operates.

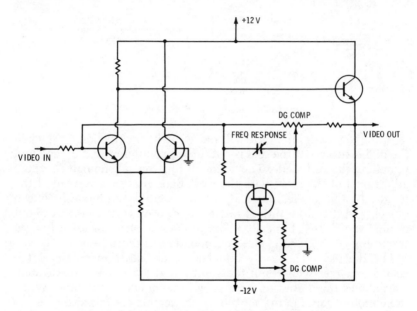

Fig. 15-27. Circuit for differential-gain predistortion.

In the shuttle mode, the tape may be rewound or moved in fast forward, depending on the direction in which the control is turned from the center position. The control can be turned about 120° in each direction, and it has seven discrete steps on each side of the center position. Each position corresponds to a definite tape speed and is indicated by a small green LED on the panel above the dial. The center position is indicated by a red LED, and this corresponds to the "still" condition of the tape. When the dial is moved to some position other than "still," it determines the speed at which the tape will move. The tape will then maintain that speed until the end of the tape is reached.

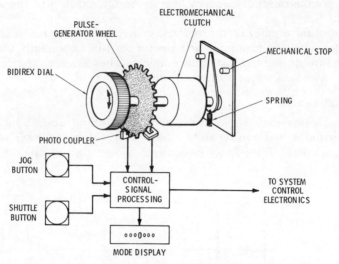

Fig. 15-28. Diagram of Bidirex control.

In the jog mode, the tape will move in either direction only when the dial is turned. If the dial is not turned continuously, the tape will remain in the still mode. The action of this control simulates manual adjustment of the reels, with the work done by servos instead of the hands. In this mode, the dial can be rotated continuously like a wheel, and it can be turned fast enough to move the tape at about normal speed. It is basically a fine position control for identifying a frame for editing or for close examination of the tape.

The dial is mechanically attached to an electromagnetic clutch that disengages when both the shuttle and jog modes are vacated. This allows the dial to turn freely with no effect on the tape. When the clutch engages in the shuttle mode, detents are felt at the center and end positions. Behind the dial is a toothed wheel, and the teeth move between two optocouplers, spaced 90° apart so that one optocoupler is off while the other is on. This allows the direction of rotation to be sensed. In the shuttle mode, as the dial is turned from the center position, an up-down counter counts the number of pulses made by the wheel as it passes the couplers. This count determines the speed at which the tape will travel. In the jog mode, the output pulses are divided by 4 and are used to pulse the capstan servo to move the tape.

The value of this control is that it allows the tape to be shuttled for editing and at the same time maintains the proper tension in the tape. This prevents the tape damage that can occur with manual shuttling.

Tape Timers

The machine has two internal tape timers that share the same display unit. The readout uses seven digits to indicate hours, minutes, seconds, and frames.

Timer 1 is normally displayed. It can be reset to zero at any time during recording or playback and is useful for timing short sections of tape. Timer 2 can be displayed only by pressing a button. This timer cannot be reset; hence, it always displays the elapsed time from the beginning of the tape. This has several production advantages in editing.

A time-code input can be fed to the cue channel, and a set of two optional boards can be included to record and read the time code. In either case, the time code can be displayed on the readout in place of Timer 1. In this case, the display uses eight digits, and thus it is easy to see whether the tape timers or the time code is being used.

Part of the timer circuit is the "zero memory" function. This automatically stops the tape in the rewind or fast-forward mode when the counter reaches the all-zeros position. If the tape is traveling at high speed, it will be slowed down as it approaches the zero position and brought to a stop gently. This permits a rapid return to the beginning of a segment for editing, etc.

The tape counter is driven by an optical wheel mounted behind the serrated tape-timer capstan, which is placed between the main capstan and the take-up reel. The optical wheel senses both the direction and rate of revolution and thus gives an accurate indication of the tape speed, position, and direction. At normal speed, the counter wheel produces 30 pulses per second, which drive standard TTL circuitry.

When the counters are used with the Bidirex dial, the tape can be positioned with one-frame accuracy. By resetting the counter to zero, an exact frame can be identified for editing, and edits can be made with one-frame precision. The counter pulses and the Bidirex function can be fed to a remote control unit that has an edit preroll function of the same accuracy.

System Control Circuits

The system control functions make up a major part of the circuitry. These are all digital circuits mounted on separate PC boards and using standard TTL ICs. They accept inputs from the operating controls and the servos and then perform a variety of functions. For example, the playback button can be pressed while the tape is in the high-speed rewind mode. However, the capstan roller will not close until the tape has been brought to a standstill, thus ensuring that the

345

tape is not damaged. Other typical functions are entering a preselected edit mode only at the correct preset time and only when color framing has been achieved, and preventing the edit mode from being entered if the servos have not locked up to the incoming video signal. Due to the variety of inputs and outputs, it is not possible to give a meaningful block diagram or full discussion here.

Remote Control

A remote-control panel that duplicates all the functions on the control panel of the machine is available. Connection to the remote control is by means of a 10-pin connector, but only one coaxial cable is used. All the functions and control lights are represented by 64 digital codes that are sent sequentially over this line at a 200-kHz rate. A "loop-through" mode in which one remote panel can control several machines is possible. With this arrangement, a master machine can be put into the playback mode and the others in the record mode so that multiple copies can be made simultaneously. Alternatively, several playbacks can be rolled at the same time as one record machine, and edits or dissolves between several playbacks can be made to build a master tape for broadcasting.

Maintenance and Service

All routine optimization of the electronics can be performed from the front panel with a small screwdriver. All the test points needed for more extensive troubleshooting have been placed on the front edge of the appropriate PC boards, along with the required controls.

Several warning and protection systems have been included. The four warning devices utilize the indicator lamps on the control panel. The indications are:

1. The STOP lamp flashes when the incoming signal or reference sync is missing.
2. The STANDBY lamp flashes when the head drum is rotating at less than $\frac{2}{3}$ of normal speed.
3. The AUTO STOP lamp flashes when the temperature of the reel motors is too high.
4. The SEARCH LED flashes when the record test switch is in the Test position. This is the red LED in the Bidirex indicator.

The purpose of these warnings is to make the operator aware of the most common problems that can cause faulty operation.

Some simple protection mechanisms are included to protect the tape and the machine from damage. These operate as follows:

1. If the still-frame mode is engaged, a timer releases the tape tension around the drum after about three minutes. This prevents head clogging and tape damage.

346

2. In the standby mode, the head is running and the tape is wrapped around its path. The tape is released after three minutes to protect the tape and the heads.

3. A tape slowdown mechanism operates as the end of the tape is reached. This usually operates in the rewind mode when the tape is being removed from the machine. It prevents damage to the end of the tape, which usually results in small pieces of tape becoming enmeshed in the heads.

Physical Layout

This machine is designed in five basic parts, which can be rack mounted individually. Fig. 15-29 shows these parts and the connecting cables. The control panels can be placed up to 200 feet away from the deck and the electronics. A compact mounting system can

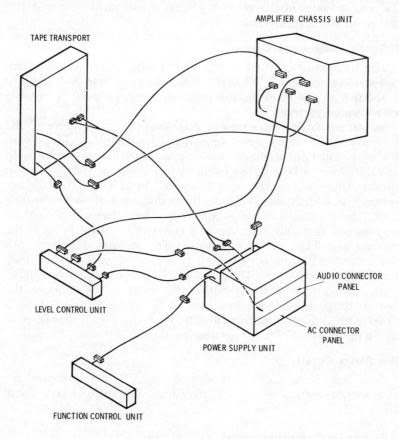

Fig. 15-29. Sections of BVH-1000 vtr.

be used in a mobile truck or a confined room, or the layout can suit the operational convenience of any size and shape of studio.

Editing

Editing is a major part of tv production, and any vtr used in broadcasting must be capable of making perfect edits in both the assemble and insert modes. In the assemble mode, the BVH-1000 lays completely new information on all tracks. In the insert mode, any combination of video, audio 1, audio 2, and cue tracks can be selected. The desired combination is preset with buttons on the front panel, and only those selected will enter the record mode. Editing can be initiated manually by pressing the EDIT button, or the remote panel or a separate computer editor can be used. A computer editor requires an interface board so that it can accept the remote codes of the vtr and so that it can provide output signals to operate the machine.

TBC and Heterodyne Option

The BVH must be used with a digital TBC to provide a broadcast-stable color signal. When it is used alone, its time-base stability is about 1 microsecond peak-to-peak, so it can produce acceptable monochrome pictures.

Sometimes this machine may be used in applications where a TBC will take up too much space or will not be necessary. To prevent having production personnel see only a monochrome picture with unstable color, a heterodyne color corrector can be included as an option. This removes the color from the direct signal and applies heterodyne techniques similar to those discussed elsewhere in this book. The result is a stable picture on a color monitor. It should be emphasized that this is a viewing convenience only to save the expense of a TBC and to save space. The pictures produced with this corrector can be copied onto cassettes for viewing, but they are not broadcast stable. The heterodyne color corrector does, however, contain a circuit for stabilizing the color on the screen in the fast shuttle modes. Color lock can be maintained up to about 7 times normal speed, and a monochrome picture will appear quite stable up to about 30 times normal speed.

The Servo Circuits

Four main servo circuits are used in the BVH-1000 to control the tape motion and to record and play back the video signal. These are:

1. The head, or drum, servo
2. The capstan servo

3. The tension servo
4. The reel servo

A logic circuit is used for the color framing, which provides inputs to the capstan servo to ensure correct color framing for edits and correctly phased playback into a studio switcher. In recording, it ensures that edits begin at the start of an odd field and finish at the end of an even field. This keeps the subcarrier continuous on the tape and avoids horizontal shifts when a digital time-base corrector is used.

For the head, main capstan, and tension capstan, pulse-driven three-phase motors are used. When used for the capstan, these give the widest dynamic range in tape speed while maintaining a steady tape flow from the slowest step motion to the maximum driven speeds. When used for the head, the three-phase motor ensures the smoothest possible running and maximum amount of control of the head position and phase. Other advantages are maximum performance, efficiency, and durability when compared to other motors of comparable size.

The first three servos are all basically the same circuit, but with different inputs for reference and feedback. The following paragraphs will look at each in general terms.

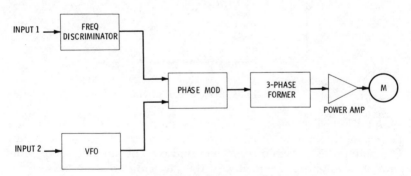

Fig. 15-30. Basic servo circuit.

Basic Servo Circuit—The basic servo circuit is shown in Fig. 15-30. The variable-frequency oscillator (vfo) produces a square wave that actually drives the motor. The output of the vfo is passed through a phase modulator to a three-phase former circuit. Here the square wave is converted to a three-phase drive with pulses 120° apart. A power amplifier raises these to the level required to drive the three-phase motor.

Two inputs are provided to the servo. Input 1 is from the FG coil in the motor. This input can be generated by a multitoothed wheel on the motor shaft, which produces a high-frequency output from an inductive pickup. This output is passed through the frequency discriminator to detect any speed changes, and the dc output from the discriminator is used to alter the duty cycle of the main drive square wave in the phase modulator. This keeps the motors running at the correct speed.

Input 2 is different for each of the servos, and will be discussed as each servo is covered. Its effect is to vary the frequency of the square wave out of the vfo.

The frequency discriminator, vfo, and phase modulator are covered briefly in the next paragraphs.

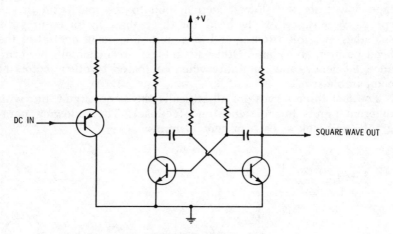

Fig. 15-31. Basic circuit of vfo.

The VFO—The basic vfo circuit is shown in Fig. 15-31. It is a simple multivibrator with the frequency controlled by a dc input. The dc input is obtained from the usual type of sampled-ramp comparator. The output frequency of the vfo is different in each of the servos, because the motors run at different speeds.

FG Discriminator—A schematic diagram of the FG discriminator is shown in Fig. 15-32. Basically, it is a phase-locked loop in which the input frequency is compared to the frequency of a locked oscillator, and the resulting control dc is used as the output. The signal from the FG coil is shaped to a square wave by Q1 and Q2. The negative-going edges from Q2 turn Q4 on, thus discharging C1. When Q4 is off, C1 charges linearly due to the constant current supplied by Q3. Transistor Q5 is an emitter follower, and Q6 is a switch. This switch is turned on by pulses derived from multivibrator Q9-Q10, and this effectively samples the ramp. Holding

350

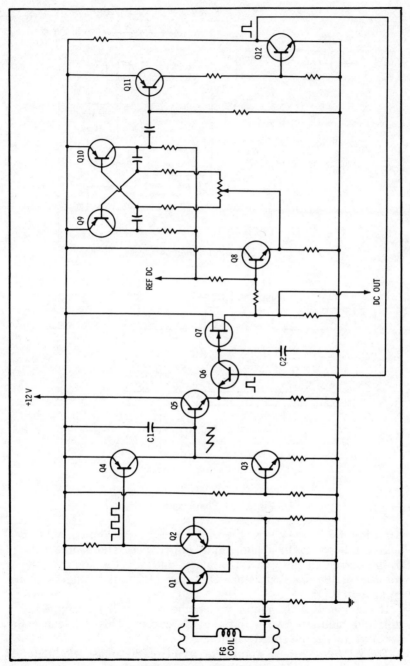

Fig. 15-32. Diagram of FG discriminator.

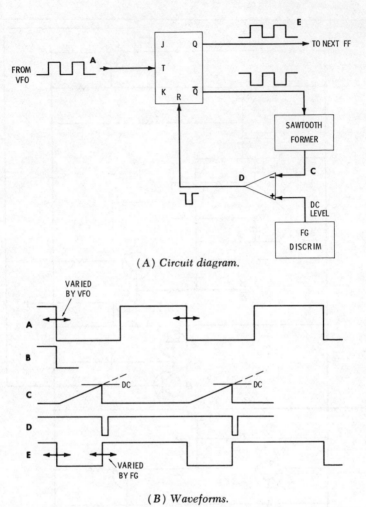

(A) Circuit diagram.

(B) Waveforms.

Fig. 15-33. Phase modulator.

capacitor C2 "remembers" the sampled dc level. Transistor Q7 is a source follower with a high input impedance. The source voltage is the output to the phase modulator and also the control voltage for the oscillator. Transistors Q11 and Q12 isolate the oscillator from switch Q6.

If the speed of the motor varies, the ramp will be sampled at a different point to give a different dc output. This will cause the servo to correct the speed variation.

Phase Modulator—A simplified circuit of the phase modulator is given in Fig. 15-33A. The output from the vfo is a square wave

352

(waveform A in Fig. 15-33B), the negative-going edge of which triggers the JK flip-flop. The Q output goes low when the flip-flop is triggered (waveform B). When this happens, the ramp at the output of the sawtooth former is started. The ramp (waveform C) is fed to the inverting input of the op-amp, which is used as a level comparator. The noninverting input is the dc level from the FG discriminator. When the ramp reaches this dc level, the output of the op-amp goes negative (waveform D), and this resets the flip-flop. The Q output then goes positive (waveform E).

When the flip-flop is reset, it resets the ramp and the output of the op-amp, so the op-amp output consists of very short pulses. If either of the incoming signals varies, the triggering and resetting of the flip-flop are altered in time. This produces an asymmetrical output waveform from the flip-flop. The leading edge is controlled by the vfo, and the trailing edge is controlled by the FG coil.

In the BVH-1000, two of these circuits are used, with the trigger input to the second being the Q output of the first. The FG-coil dc level is fed to both. Two circuits are needed because one does not give a wide enough range of control for the motors.

Head Servo—In the head servo, the head drum provides two feedback signals: one PG pulse per revolution, and a 6293-Hz signal from the FG generator. A diagram of the servo is shown in Fig. 15-34. The servo goes through three stages of lock before reaching its final operating state. These stages are:

1. The PG pulses are compared to reference vertical in phase comparator 1. The PG pulses form the sampling pulse, and the reference vertical forms the ramp. When lockup is achieved, lock detector 1 substitutes playback vertical for the PG pulses.
2. The playback vertical pulses are now compared to the reference vertical, again in phase comparator 1. When lock is achieved, lock detector 2 engages phase comparator 2.
3. Now playback horizontal and reference horizontal are compared in phase comparator 2. When these are locked up, the head is under the tightest servo control, and the output can be phased and mixed with other sources (provided a TBC is used). When this final lock is achieved, the outputs of lock detectors 2 and 3 enable the AND gate, which illuminates the front-panel horizontal-lock indicator light.

The vertical-lock signals control a variable multivibrator. This is used to control the pulses out of the free-running vfo and into the phase modulator to achieve a quick lockup. When the switch to horizontal lock occurs, the input is directly to the vfo. The reference horizontal and vertical signals are obtained from the external sync or advance sync input.

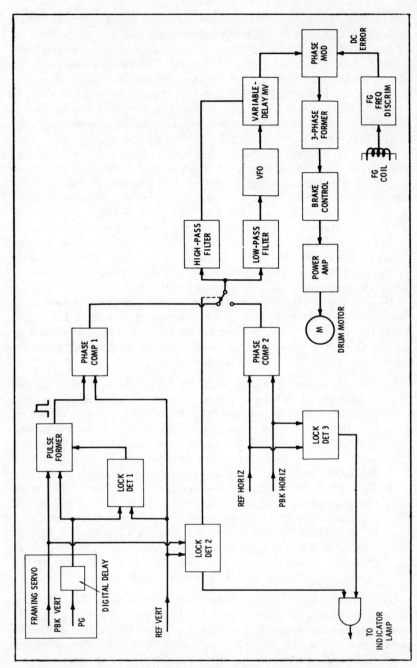

Fig. 15-34. Block diagram of head-drum servo.

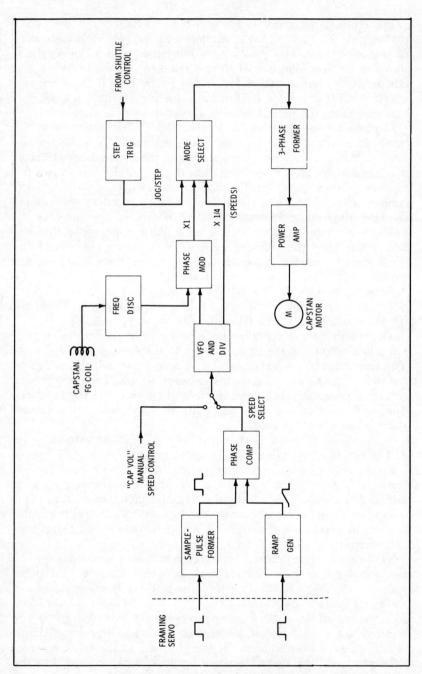

Fig. 15-35. Block diagram of capstan servo.

355

For editing, the servo locks onto the PG tach pulse through a digital phase shifter, which is itself comparing playback vertical to reference vertical. This produces a delay that phases the playback signal to the incoming signal so that edits can be made. When the edit is made, the digital delay is "memorized" and then remains constant. This ensures a continuity of picture on playback. In normal recording, the digital delay is preset to a nominal value.

Capstan Servo—The capstan servo (Fig. 15-35) is similar to the basic circuit, but it has two modifications added. The first is the mode-select circuit. This is used to change the motor drive so that it is controlled by the shuttle control in the jog mode, and to run the tape in the low-speed shuttle modes. The second is the "cap vol" control. This is a manual speed control that is used to synchronize the tape playback to a network feed or to another tape machine.

The other input to the vfo is from a phase comparator that has two inputs from the framing servo. These two inputs control the three framing modes in which the capstan can work. These are explained in connection with the framing servo.

Framing Servo—The BVH-1000 can operate with three different degrees of servo lock with respect to the incoming signal. These are the field, frame, and color-frame modes.

The framing servo contains the logic for these modes, and it provides two outputs to the capstan servo to achieve the required lock. The two outputs are a ramp and a sample pulse, which are fed to the phase comparator in the capstan servo circuit. The three modes represent progressively tighter degrees of control over the tape motion, and each is necessary for edits that are free of horizontal shifts when a TBC is used.

A simplified diagram is shown in Fig. 15-36. The external sync and subcarrier produce a 15-Hz color-framing pulse, 200 microseconds wide, which identifies every fourth field. Reference vertical forms a 30-Hz square wave to which the color-framing pulse is added 200 microseconds after the negative-going transistion. This composite signal is recorded onto the tape on the control track and is used in playback to identify the field sequence and to achieve color framing.

This signal is also passed into a circuit that is a combination of monostables and gates. This circuit produces three output pulse trains at 15-, 30-, and 60-Hz rates. One of these is selected to form the ramp in the capstan servo.

In the record mode, the capstan vfo output is divided in frequency and fed to a similar circuit, which also produces three pulse trains. One of these is used to form the sample pulse in the capstan servo.

In playback, the same reference vertical signal is used, but the vfo output is replaced by the playback control pulses.

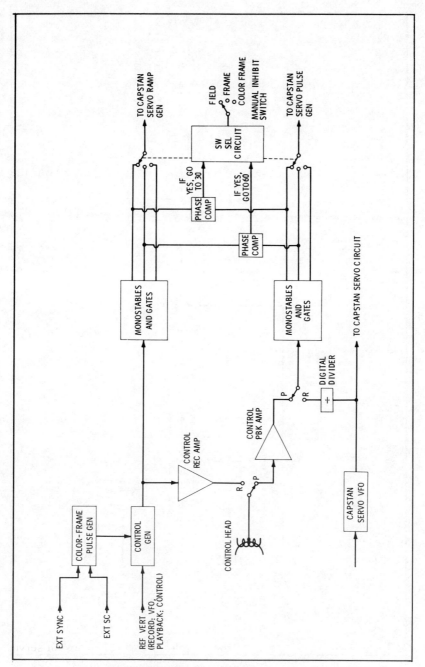

Fig. 15-36. Block diagram of framing servo.

357

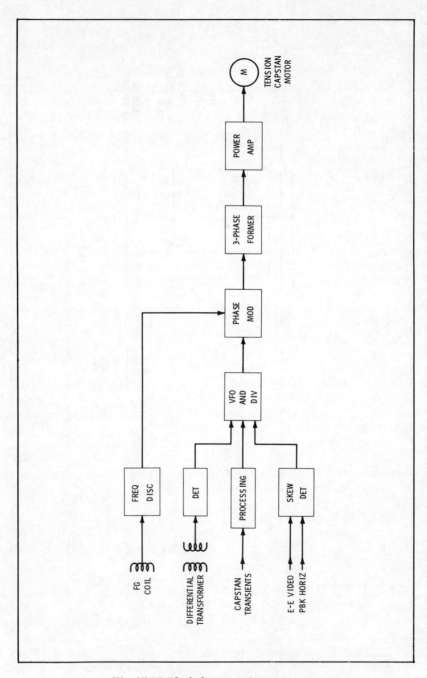

Fig. 15-37. Block diagram of tension servo.

The three output pulses from each circuit are fed to a three-position switch. This is an electronic switch that starts in the 15-Hz position. A phase comparator compares the two 15-Hz waveforms, which are gradually brought into coincidence as the capstan runs up to speed and pulls into the correct phase alignment. When coincidence is achieved, the switch moves to the 30-Hz position. When the capstan locks up to these pulses, the switch moves to the 60-Hz position, to produce the tightest servo control and hold the servo in the color-frame mode.

A manual three-position switch can inhibit the action of the phase-comparator switching in either the field, frame, or color-frame mode if this is required.

To review the inputs and the operational modes: In the record mode, the capstan vfo output is compared to the incoming video to obtain the correct phase of the capstan. The actual signals are the reference vertical and the vfo output. In playback, the reference vertical is compared to the control-track pulses. The servo locks to the color-frame pulse, so all edits will have the correct color phasing to prevent horizontal shifts when a digital TBC is used.

Tension Servo—Tape tension is vitally important in a vtr, especially in a full-wrap helical machine. To help the precision tape guiding system in the play and record modes, a tape-drive servo system is used. It uses two capstans, the main drive capstan and the tension or feed capstan.

In the record mode, the main capstan rotates at a fixed speed, and the tension around the drum is controlled by the feed capstan. A movable guide, which is coupled to a differential transformer (Fig. 15-37), contacts the tape. A 2-kHz signal is fed into the transformer, and variations in the tape tension move the guide, alter the coupling, and thus alter the output amplitude of the 2-kHz signal. This signal acts as one feedback input to control the vfo, which adjusts the motor to control the tape tension.

In playback, two sets of horizontal-sync pulses are compared. These are taken from the tape as it enters the drum and as it leaves the drum, that is, just before and just after the switching point. An error voltage is generated, and this voltage controls the tape tension to give minimum skew of around 1 to 5 microseconds.

To prevent tape stretch with the dual capstan, an input from the main capstan servo is also fed to the vfo, so that the small changes in capstan speed are reflected in the feed capstan. To prevent the tape from stretching and binding around the drum at start-up, the main capstan closes just after the feed capstan closes.

This servo controls the tension of the tape between the two capstans and around the drum only. It does not control the torque in the two reels or affect the tape outside the head-drum loop.

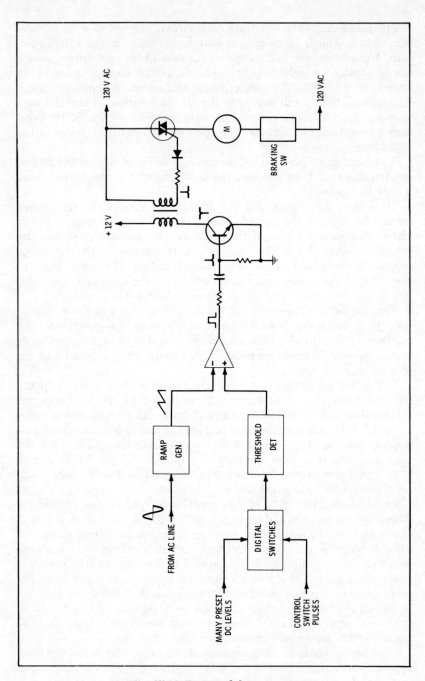

Fig. 15-38. Basic reel-drive circuit.

360

Reel Servos—The reel servos are very complex and can be covered only lightly here. The two reels are directly driven by single-phase 120-volt ac motors. The speed and torque of these motors are controlled by triac circuits in the various drive modes, but in the braking mode the drive is modified slightly. Fig. 15-38 shows the basic drive circuit.

A ramp is formed from the zero crossing of the power-line voltage and is fed to the inverting input of the op-amp. The noninverting input is a dc level from the threshold detector; this will be covered later. The dc sets the level on the ramp at which the output from the op-amp goes positive (Fig. 15-39). The op-amp output

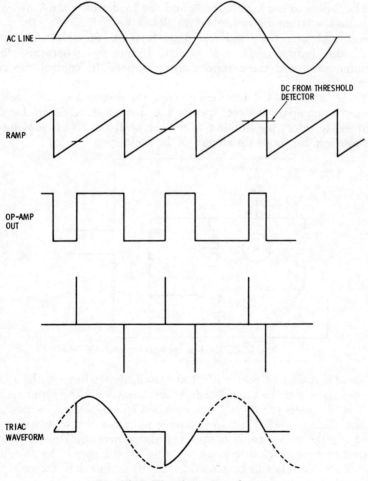

Fig. 15-39. Waveforms in reel servo.

is differentiated, and the positive-going edge turns on the transistor. The collector pulse from the transistor fires the triac through the transformer. As the dc level varies, the pulse width out of the op-amp varies, and so does the position of the positive pulse with respect to the input power-line sine wave. Hence the triac is triggered at different points on the sine wave to produce pulses of varying power to the motor. (This action is similar to that of a light-dimmer circuit.)

The main part of the servo logic is located before the threshold detector. The input to the threshold detector is one of many dc levels that are set by bias resistors. These are chosen by a series of electronic switches that are controlled by the servo logic circuits.

The inputs to the logic circuits and the modes of control are complex and will not be described in detail here. Basically, the tape speed, tape direction, and reel diameters are all sensed as inputs from the Bidirex dial, reel optical frequency generators, tape counter, etc., and these inputs are processed to control the reel motors.

There is an optical frequency-generator sensor for each reel. It is used to compare the reel speeds and diameters, to determine the torques, to select the braking, and to sense if the reel is moving or not moving. It is not a direct part of the reel servo.

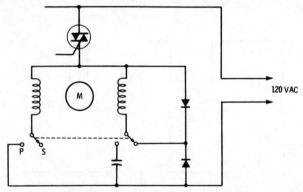

Fig. 15-40. Reel-motor braking control.

Various modes of reel control are used, depending on the mode of operation and the tape speed. As an example, in the rapid search modes, the reels are driven at a constant speed, set by the position of the Bidirex dial. The tape counter monitors the tape speed by counting the pulses from its optical pulse generator. These are converted to a dc level that is proportional to the speed—by selecting one of the switches to be closed in the threshold detector—and the selected dc controls the reel speed.

Another example is provided by the braking action. Braking depends on the amount of tape on the reels, since more tape means a higher inertia of the reel. Two modes of braking are used. Above six times normal speed, the ac feed to the motor is half-wave rectified and fed to one coil only, which now acts as an eddy-current brake (Fig. 15-40). When the tape slows to less than six times normal speed, the drive is returned to normal, and the back tension and motor torque slow the tape.

When the tape is not in motion, parking brakes are applied by solenoid action. This is the only time these brakes are used; they are not used as friction brakes during tape motion.

One important point is that the tension servo does not provide an input to the reel servos. In the record and playback modes, tape tension outside the closed path around the head drum is controlled by the reel servos, and the inputs from the optical sensors are used to keep the tension within limits. In the shuttle modes, the tape tension is, of course, controlled by the reel torques.

16

The Portable VTR

Several manufacturers have a portable vtr in their line of products, and the facilities possessed by all are similar. The idea of a portable vtr is to allow a reporter to video record program or news material in the field without the need for a power line or other limitations or restrictions. The vtr deck and camera are made as a set of two separate units connected together by a single multiconductor cable. Fig. 16-1 shows typical modern portable vtr's and cameras.

The older type of camera had a rifle-type sight and an external mounting for the microphone. The newer models have a built-in electronic viewfinder, and the microphone is mounted above the zoom lens as an integral part of the front plate.

The vtr deck is provided with a leather carrying case and shoulder strap and is made so that it can work in either the vertical or horizontal position. A typical deck weighs about 20 pounds, and a typical camera weighs 5 or 6 pounds.

Power is provided by a sealed rechargeable battery mounted inside the deck. This will provide power for about 45 minutes to one hour, allowing at least two 20-minute reels to be recorded and rewound. Indoor operation can be extended by the use of a separate battery-eliminator pack that plugs into an ac outlet and provides the required power input to the deck and camera. This accessory will also charge the battery, but individual models should be checked to see if the vtr will run and the battery will charge simultaneously.

Due to the necessarily compact nature and small size of the portable vtr, the path of the tape tends to be a little more complicated than in the larger machines, as is shown in Fig. 16-2. However, the wrap around the head drum is the same.

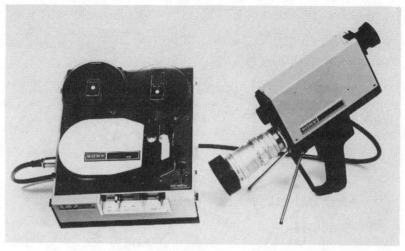

Courtesy Sony Corp. of America

(A) *Sony.*

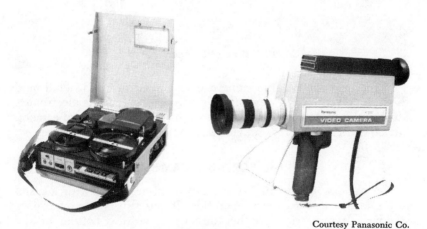

(B) *Panasonic.*

Courtesy Panasonic Co.

Fig. 16-1. Examples of portable vtr's.

An input to these machines can be provided from a video monitor or a tv output to allow normal vtr recording. The early models could only record from their own camera and did not have playback capability. The tape had to be placed on a compatible machine for viewing. The later machines can play back either into their own camera viewfinder or into a monitor or tv set, through the multiconductor cable or an rf unit inside the deck.

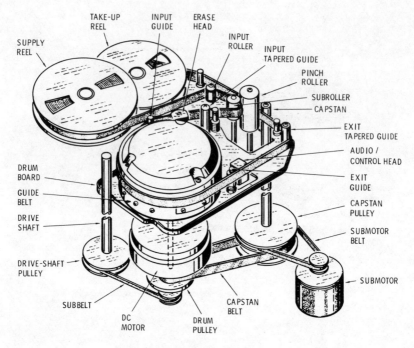

Fig. 16-2. Tape path in portable vtr.

In the next section, the camera is examined in more detail, and this is followed by a review of a typical vtr deck and its electronics. Most of this material is based on the Sony AV3400 deck and camera, but other manufacturers provide similar machines and facilities.

THE PORTABLE CAMERA

The camera in Fig. 16-3 is a small hand-held unit with a pistol grip and trigger for ease of operation. It can also be mounted on a standard 16-mm tripod, either directly or at the bottom of the handle. Fig. 16-3 points out the salient operational features, and it should be noted that all the electronic controls are internal and are not accessible to the operator.

The lens is a zoom lens, which can have a focal-length variation of from 4:1 to 10:1, depending on the model used. Both focusing and iris controls are provided on the lens, and the standard 16-mm "C" mounting has an adjustment for setting up the zoom.

During recording, the picture can be viewed and framed with the small viewfinder, which is in fact a miniature tv screen. A red light mounted inside the eyepiece indicates recording is in process. If

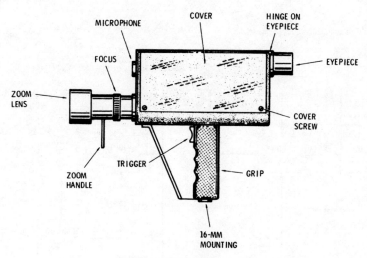

Fig. 16-3. Portable camera.

the tape runs out or the machine is stopped, the power to the viewfinder is interrupted, and the raster goes out.

In the record mode, the video signal from the vidicon is fed to both the vtr deck and the viewfinder. The sync and scanning signals are derived in the vtr deck and fed to both the vidicon and the viewfinder. On playback, the video signal from the tape is fed to the viewfinder, thus providing a useful field check that a recording has been made. The video from the tape is also stripped of its sync signal, and this is used to feed separate horizontal and vertical drives to the viewfinder. On both record and playback, a 2:1 interlace is maintained, but the viewfinder scans are only frequency locked to the playback and are not phase locked. This means that on playback the picture in the viewfinder can be offset to one side and not framed exactly as it was in recording.

A supply of 12 volts is fed from the deck to the camera, where it is regulated down to 9 volts to drive the camera circuitry. This voltage is provided at all times, and so the vidicon is powered even on playback. This permits a recording to be made immediately after a playback without a warm-up period.

A 3-kilovolt supply for the viewfinder is derived from the incoming horizontal pulses, and thus the viewfinder loses its raster when the tape is stopped. If the 9 volts is off slightly, the viewfinder will be out of focus, and this is a good indication that the battery is beginning to run low.

Fig. 16-4 is a block diagram of a typical camera, and Fig. 16-5 explains how the viewfinder is fed the playback video signal instead

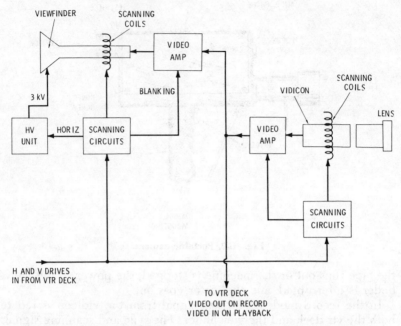

Fig. 16-4. Block diagram of portable camera.

of the vidicon signal without the use of a mechanical switch. The incoming video is riding on about a 5-volt positive dc, which back-biases the emitter of Q1, thus blocking the vidicon signal. Removal of the playback signal allows normal emitter-follower operation of this transistor.

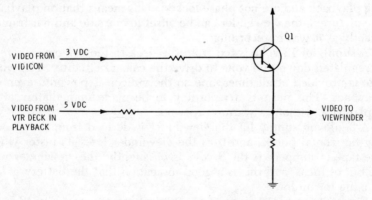

Fig. 16-5. Video switching circuit.

THE PORTABLE DECK

The portable vtr deck can be divided into three distinct sections, the mechanical construction, the video circuits, and the servo circuits. Each will be covered separately in the following subsections.

Mechanical Construction

Due to the requirement of portability, this machine must produce a usable length of program material in a standard format, and yet contain everything necessary in the smallest, lightest, and most compact format possible. This is achieved by using smaller reels and an unusual tape path. To conserve space, all the required levers and plugs are placed on the front and side of the machine and are made readily accessible (Fig. 16-6).

The head drum is the same as that for larger machines of the same format, and it has the same heads, which are serviced and

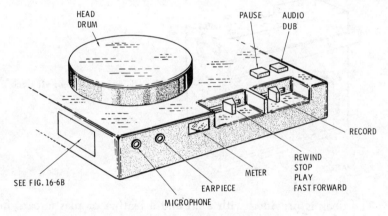

(A) *Locations of controls and connectors.*

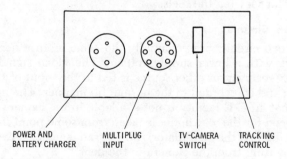

(B) *Additional controls and connectors.*

Fig. 16-6. Portable vtr deck.

removed in the same manner. All of the electronics and motors are packed inside the deck in a highly compact fashion.

The battery and the rf unit are inserted and removed through a special door in the bottom of the case, as shown in Fig. 16-7. They are held in place by the mechanical design and not by any screw-down device or other type of fixture.

The main connection from the deck to the camera or a tv set is a multiconductor cable, which carries all of the signals and power required. Although all manufacturers use a similar plug and cable for this purpose, the connections can be considerably different and should always be checked when it is necessary to mix equipment.

All of the motors used are low-voltage dc types, and a mixture of direct and belt drives is employed. Fast forward and rewind are accomplished through mechanical drives powered from the main head motor.

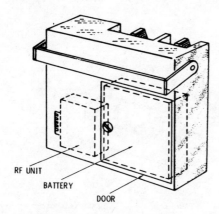

RF UNIT

BATTERY

DOOR

Fig. 16-7. Underside of vtr deck.

The deck is provided with a lid and a leather or plastic carrying case, both having cutaway sections for the controls and transparent windows for viewing the reels.

The Video Circuits

The video circuits are essentially the same as in other vtr's, but they work with a lower supply voltage. The video signal from the camera, tv receiver, or other source is fed to the input of the recording chain and is recorded in the normal fm manner. The main difference is that an E-E facility is not available in the camera mode. Instead, a feed for the viewfinder is taken from some point in the video amplifier before the modulator. This arrangement shows that the signal is getting from the camera to the deck.

Playback is exactly as in any other vtr. The full video signal is recovered inside the machine and appears at the output, where it is fed to the camera viewfinder or a tv monitor. This is a full video signal with sync and is also fed to the rf modulator. The modulator output is on a regular tv channel that is vacant in the area in which the vtr is used.

The Servo Circuits

The portable vtr differs from other machines in that its prime function is to record the output from a camera without the usual 60-Hz reference derived from the ac line or the incoming video signal. Normally, the incoming signal provides the reference vertical sync to which the vtr servo locks; thus the vtr in effect slaves to the incoming vertical sync.

In the portable vtr, the reverse is true: the vtr actually generates the reference pulses, and these drive the camera. The rotating heads are run at an exact speed, and their tachometer pulses are the reference to which everything else locks. The head rotation produces the

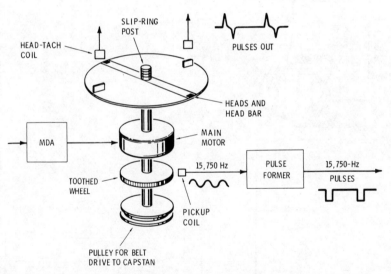

Fig. 16-8. Pulse generation in portable vtr deck.

vertical and horizontal drives to the camera as well as the control track for the tape. This new mode of operation accounts for the slight differences of servo operation that are found in the portable machines.

Both head and capstan servos are used, and each mode of operation of the machine uses these servos slightly differently. The following sections cover this in more detail.

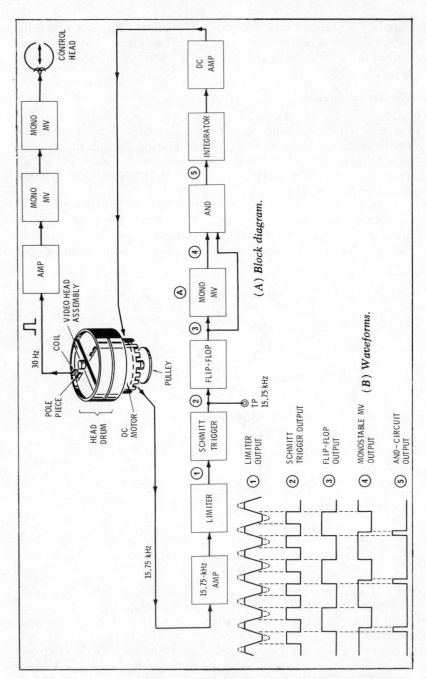

Fig. 16-9. Camera record servo.

(A) *Block diagram.*

(B) *Waveforms.*

① LIMITER OUTPUT

② SCHMITT TRIGGER OUTPUT

③ FLIP-FLOP OUTPUT

④ MONOSTABLE MV OUTPUT

⑤ AND-CIRCUIT OUTPUT

THE SERVO IN RECORDING

In the record mode, there are two different modes of operation. These are recording from the portable camera and recording from a tv receiver or other source.

Recording From a Portable Camera

When the machine is recording from a portable camera, the head servo is set so that the heads free-run to produce a stable 15,750-Hz signal from the frequency-generator coil in the head drum. The capstan servo is disabled, and the tape is run past the heads at a constant speed.

Fig. 16-8 shows the basic layout of the head drum and drive mechanism, with the signals produced by the head-tach coils and the frequency-generator coil. The vertical and horizontal drives for the camera are taken from these coils, and thus the camera is synchronized with the rotating heads by a 2:1 interlaced drive. The occurrence of the vertical pulse is controlled by the head position, so the vertical pulse always appears when the head is in the correct position to record it. In this way, the video tracks on the tape are always correct. The vertical-drive pulse from the head-tach coil is also used for head switching.

Fig. 16-9A is a block diagram of the servo system in the record mode. Speed changes of the head are sensed by a coil under the head assembly, which produces a 15,750-Hz signal. This is amplified, limited, and used to drive a Schmitt trigger circuit. The Schmitt trigger produces a clean pulse at the same frequency (waveform 2 in Fig. 16-9B), and this pulse is used to change the state of a bistable flip-flop. Thus the on time of the flip-flop is controlled by the rotational speed of the heads.

The negative slope of the flip-flop pulse triggers the positive edge of the output of the following monostable multivibrator (A in Fig. 16-9A), which is the heart of the servo system. Its negative slope is set by a potentiometer, and thus its on time is held constant.

The pulse lengths of the flip-flop and monostable A are compared in a gate, the output of which is a narrow pulse with its width dependent on the widths of the two input pulses. Thus variations in the speed of the head vary the width of the output pulse from the flip-flop, which thus varies the width of the output pulse from the gate. This output pulse is integrated to a dc level, which is further amplified and used to drive the dc motor.

In setting up the vtr for correct servo operation, it is vitally important to set the width of the output pulse of monostable A. This is performed with a frequency counter connected to the output of the head coil, and the potentiometer is adjusted until 15,750 Hz is read. One end of this potentiometer is connected to the supply voltage, and this must first be set accurately. A setting of the power-supply voltage at the regulator that is one volt high will lower the free-running frequency of the heads by as much as 400 Hz. A seri-ously low battery will have the same effect. The result of this is that the recordings are unplayable on any other machine because the heads were running at completely the wrong speed.

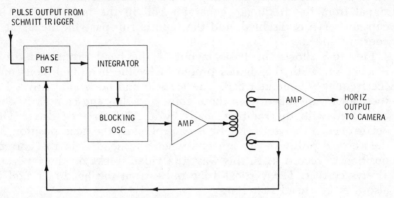

Fig. 16-10. Horizontal-pulse control servo.

The output of the Schmitt trigger is also fed to a phase compara-tor (Fig. 16-10), where it is compared with the horizontal output to the camera. The error signal developed controls the frequency of a blocking oscillator, the output of which actually forms the horizontal drive. In this way, the horizontal pulses are servo locked to the heads at all times.

The capstan is not servo controlled, but is run at a constant speed by the belt drive from the main head motor. The variations in the head speed are so slight compared to the capstan speed that they are swamped completely by the belt drive, and the capstan drives the tape at an even speed.

Recording From a TV Receiver or Other Source

When recording from a source other than the camera, the port-able vtr operates the same as any other vtr. The heads are servo controlled, with the incoming vertical-sync pulse used as the refer-ence and the output of the head-tach coils as the feedback informa-tion. Fig. 16-11 is a block diagram of the servo in this mode. The arrangement is quite similar to that for the camera record mode.

The incoming vertical sync is processed to produce a pulse, which is then integrated to a ramp. This ramp is sampled by the head-

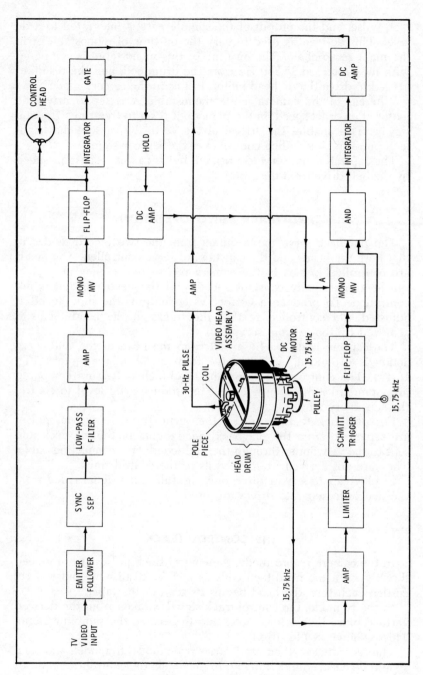

Fig. 16-11. Diagram of servo in tv record mode.

tach pulse, and the output of the sample gate is integrated to a dc level. This dc level is used to vary the on time of monostable A in the main servo chain. The on time of this monostable is compared with the processed 15,750 Hz from the drum coil, and the resultant is used to drive the dc head motor, just as in the camera mode.

Whereas in the camera mode monostable A is set to drive the motor at an exact speed, in the tv mode it is set to free-run a fraction slower. This enables the arrival of the vertical sync and the head-tach pulse to effect a fine control over the servo system.

The capstan is not servo controlled, but is run at a constant speed by the belt drive from the motor.

PLAYBACK SERVO OPERATION

The playback servo mode differs from the two record modes in that both the heads and the capstan are servo controlled. The heads are controlled just as in the camera record mode; they are set to run at a continually constant speed, and the head position is not controlled. To produce a correct video output, the tape speed is adjusted and controlled so that the tracks fit the heads. This is achieved by the capstan servo.

The capstan is driven by a belt from the head motor, and servo action is produced by belt drive from a subsidiary motor (Fig. 16-12). The control pulses and the head-tach pulses are compared in a standard ramp-sample circuit to produce a dc level to control the subsidiary motor.

For playback to the camera, the vertical and horizontal pulses are separated from the playback video signal in the vtr deck and sent to the viewfinder through the multiconductor connecting cable. They are not derived from the coils or the control track.

In playback to a tv monitor, only the full video signal is fed from the deck. No separate drives are used.

THE CONTROL TRACK

In the camera record mode, the control-track pulses are produced directly from the head-tach coils as in Fig. 16-13. One pulse is recorded each time the head begins its scan of the tape.

In the tv mode, the control-track signal is taken from the derived vertical pulse before it is integrated to become the sampling ramp. This is shown in Fig. 16-11.

The recording method used varies from model to model, and each should be checked individually. In playback, the control-track pulses are used to control the servo as already described.

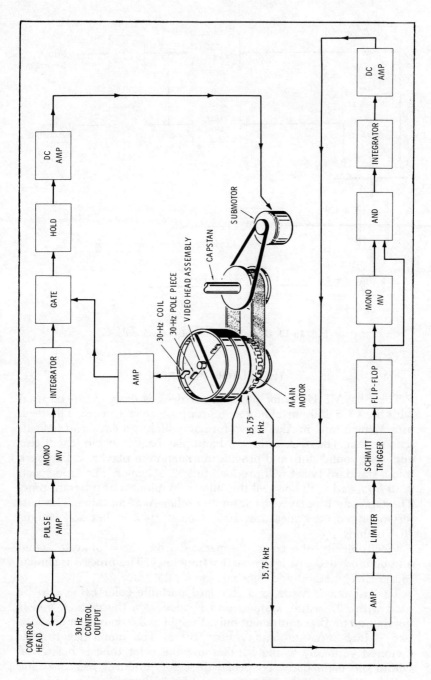

Fig. 16-12. Diagram of servo in playback mode.

377

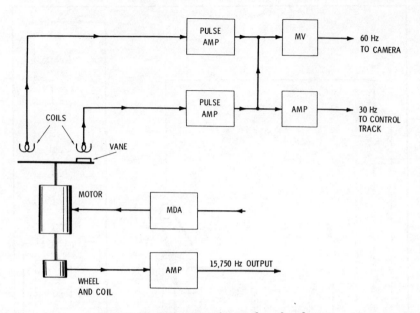

Fig. 16-13. Generation of control-track pulses.

THE COLOR PORTABLE

The Akai VT-110 color portable vtr deck is designed to be used with the CCS 150 portable general-purpose color camera. The deck uses ¼-inch tape in Akai's own format, which is a two-head helical-scan system. The deck will record and play back in color; it will also auto edit, sound dub, and provide stop action in playback. Playback into a standard tv set will produce full NTSC color. The tape speed is 10 in/s, and with one reel this allows 26 minutes of program time. The tape can be played back on any other Akai machine. The battery-operated deck measures 10.5 × 5.6 × 14.2 inches and weighs 16.4 pounds.

Electronically, the color is separated from the composite signal and heterodyned to a lower carrier frequency. The process is similar to the EIAJ method and does not use a pilot track.

The camera is made as a standard portable color camera to be used with all existing systems and vtr decks on the market, and is not limited to Akai equipment only. Weighing 5.5 pounds, it contains two ⅔-inch separate-mesh vidicon tubes. The luminance tube is mounted vertically in the handle, and the color tube is horizontal within the main body. A special dissector produces red and blue only for the color tube. This arrangement is shown in Fig. 16-14.

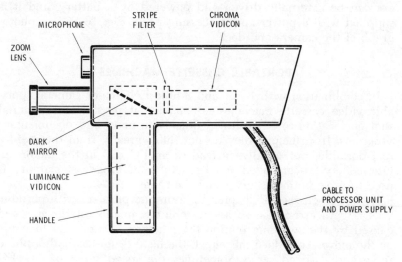

Fig. 16-14. Portable color camera.

A 6:1 zoom lens with an automatic servo-controlled iris is provided in a standard "C" mount. This has a 33:1 sensitivity range. Above the lens, a 600-ohm unidirectional microphone is built in. The cam-

Courtesy Akai America, Ltd.
Fig. 16-15. Portable color camera and vtr.

379

era can be externally driven and powered by a battery, and it is supplied with a power and processing pack. Fig. 16-15 is a photograph of the camera and deck.

PORTABLE CASSETTE MACHINES

In combination with 1-, 2-, and 3-tube ENG color cameras, portable video cassette machines are making it possible for material shot on location to be included quickly in news and documentary programs. In principle, they do not differ greatly from the reel-to-reel portable, but usually, instead of an FG coil in the motor, an internal crystal-controlled sync generator of broadcast stability is used to drive both the servos and the cameras.

As was discussed in Chapter 13, a unique problem with portable vtr's is gyro errors. These are instabilities in the rotational speed caused by the swinging motion the machine is subjected to while on the move. To reduce this effect, the head drum has both a phase servo and a speed servo. Sometimes the speed servo has an FG coil, but in many cases multiple PG coils and tach vanes are used. These produce a fairly high-frequency output and are used in the type of speed servo described in Chapter 8. This type of servo reduces the gyro errors to the point where they can be handled satisfactorily by a TBC with a gyro adapter.

The smaller machines can record only, or possibly play back a low-resolution monochrome picture into a camera viewfinder to check that a recording has been made. Fig. 16-16 shows a portable U-matic deck.

Courtesy Sony Corp. of America

Fig. 16-16. Portable cassette vtr.

Courtesy Sony Corp. of America
Fig. 16-17. Portable cassette vtr and color camera.

The main use of small portable machines is to allow a reporter to reach a scene quickly and unencumbered with heavy equipment. Ideally, the tape will be played back later at the studio on a different machine through a TBC. Fig. 16-17 shows a portable Betamax deck with a color camera.

PORTABLE BROADCAST HELICAL MACHINES

Portable models of all the major segmented and nonsegmented broadcast helical machines have been made. These are not ENG machines; they are intended primarily for use in mobile teleproduction, typically mounted in a truck or a van. They can be removed from the truck and placed on the ground, but they are not intended to be carried around like the cassette machines. Power is provided from external batteries or other sources rather than from batteries inside the machine.

A full range of operational facilities is provided. These include access to all three audio tracks with microphone or line inputs, with level meters and gain controls. All the edit functions are provided, but automatic editing usually is not. The time code is most often provided from an external portable generator rather than from inside the machine.

To ensure stability of the tape drive, separate capstan and reel motors are used, and these are usually small printed-circuit dc motors. The servos are often simpler than in the larger machines, but they are effective enough to produce stable recordings that can

381

be played back later through a TBC. Servo-lock lights are included on most models to ensure proper lockup at all times.

In most cases, the tape is intended to be played back on another machine in the studio, so usually only a low-resolution monochrome picture is provided for monitoring purposes. Often the playback electronics consists of a simple amplifier and limiter, followed by a demodulator, with no video correction added.

In the BCN portable, the deck is the same as the studio deck (except for the reel mounting), and when it is returned to the studio it can be plugged into the same electronics unit used by the studio decks. One advantage of this arrangement is that the tape can be played back on the deck that recorded it, thus removing a source of possible video timing and tracking errors. Another advantage is that several decks may be purchased, and they can all share the same electronics unit.

CONCLUSION

The original portable vtr's were simple monochrome machines. They had no playback capability, and the tape had to be removed to another machine for viewing. But, they proved the concept of mobile video to be workable. With the introduction of integrated circuits and improved circuit design, they have maintained their light weight and small size while adding many of the facilities found on the larger machines.

Because of the importance of cassette machines as "roving reporters" in the ENG field, their use will continue to grow as smaller cameras become available. The open-reel machines have changed the character of location shooting for documentaries and features, and the portable vtr is now a major piece of equipment in tv throughout the world.

17

Introduction to Digital TV

Digitizing the tv signal was first seriously proposed in the early 1950s, but although the mathematics and the requirements were fully worked out, the technology did not exist even to attempt it until the late 1950s and early 1960s. Two early examples were the Bell Picturephone and certain satellite pictures of the Earth. Both of these were slow-scan, low-resolution pictures, entirely unsuitable for broadcasting.

Production of a digitized broadcast tv signal did not occur until around 1970 when LSI semiconductor computer memories and high-speed analogue-to-digital (a/d) converters became available. This new technology made possible the development of the digital systems scan converter and the commercial digital time-base corrector.

The potential uses for digital tv are so enormous that an immediate change began to appear in tv equipment. The basic elements of digital tv will be briefly explained in this chapter, because vtr's and TBCs are directly affected by these new techniques.

DIGITIZING AN ANALOG SIGNAL

Digitizing a varying electronic signal, such as a tv or audio signal, means converting it into a coded series of discrete quantities, usually represented by the binary digits 1 and 0. Many techniques exist for this; some are fast and others are slow; some are accurate and others are not. For tv, a very fast and very accurate conversion is needed, and this is the main reason why the tv signal was the last of the commonly used signals to which digital techniques were applied.

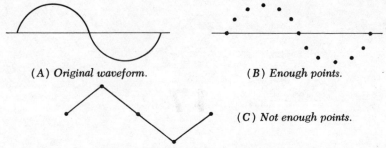

(A) Original waveform.   (B) Enough points.

(C) Not enough points.

Fig. 17-1. Representation of a waveform by points.

The easiest way to explailn a/d conversion is to start with a simple analog signal, such as the sine wave in Fig. 17-1A. This waveform can be copied by plotting points, as in Fig. 17-1B, and then drawing a curve through the points. If there are enough points, the curve will be quite accurate; if there are not enough points, a wrong curve will result, as shown in Fig. 17-1C.

Analog-to-digital conversion is the reverse of drawing a curve through plotted points. It is a process of analyzing the curve and picking points that can later be used to redraw the curve. The main problem is not where to pick the points, but how many to pick. Nyquist's theorem tells how many points are needed to plot the curve properly. It is usually quoted in terms of frequency. It says that the *sampling frequency*—a measurement of how many points are picked—must be at least twice the frequency of the waveshape being sampled. In practice this is a little low, and usually a factor of three or four times is used as a minimum.

The reason for digitizing an analogue signal is that a digital signal has several advantages over an analog signal. Some of these advantages are:

1. Digital signals are less susceptible to noise and distortion.
2. They are easy to transmit from place to place.
3. They are easy to store in a memory.
4. Digital equipment is simpler than analog equipment. It has fewer operator controls and requires less adjustment.
5. Digital signals are easy to manipulate, especially with computer-type programs.
6. Errors in digital signals can be detected and removed easily.

ANALOG-TO-DIGITAL CONVERSION

To digitize a signal, an a/d converter is used. Many types exist, but only the basic principles of those used in tv will be covered. There are three stages in a/d conversion:

1. Sampling. The analog signal is turned into a series of sample pulses.
2. Quantizing. The sampled signal is adjusted so that the sample pulses are made equal to reference voltage levels.
3. Encoding. The quantized samples are turned into a digital code.

Each of these stages will be covered below.

Sampling

The analog signal cannot be applied directly to a circuit that will generate the digital numbers that correspond to its voltage level. This is because the input signal is continually changing, and a definite number would not be produced. The original signal must be sampled at some point, and that sample must be "remembered" for a short time while the digital number is generated.

In an electronic a/d converter, the input waveform is sampled by a narrow pulse in a manner similar to that used in a vtr servo circuit. The output of the sample gate is a narrow pulse with an amplitude equal to the amplitude of the input signal at the sample point. Fig. 17-2B shows the resulting waveform (waveform B). This is often called a *pulse-amplitude modulated (pam)* waveform.

Each of the pulses is held in a *holding capacitor,* and the actual waveform becomes a series of steps (waveform C in Fig. 17-3B). This gives the rest of the circuit time to produce the binary numbers.

Quantizing

In an a/d process, the total range of the input signal must have upper and lower limits set, and the number of equal discrete steps

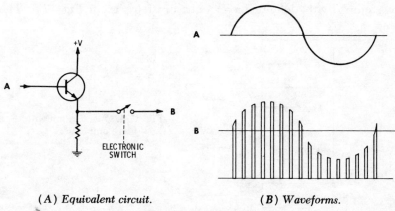

(A) *Equivalent circuit.*   (B) *Waveforms.*

Fig. 17-2. Sampling of a waveform.

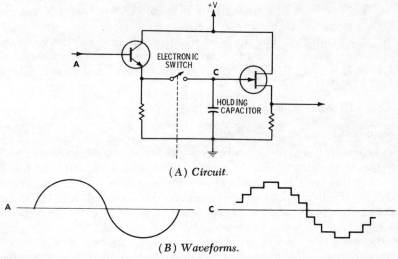

(A) Circuit.

(B) Waveforms.

Fig. 17-3. Action of holding capacitor.

between these limits must also be decided. A separate digital number is assigned to each step. The more steps there are between the limits, the more accurate the digitizing process will be.

The discrete steps are set electronically by a series of reference voltage levels, and the incoming step waveform is compared to these levels. Fig. 17-4 shows a simple circuit with only four levels. As the input signal exceeds each reference level, the output of the associated comparator goes high. Thus the pattern of ones and zeros on the four output lines shows how high the voltage level of the input signal is.

In effect, the first step waveform has been adjusted to a step waveform with definite increments, or steps, as in Fig 17-5. The

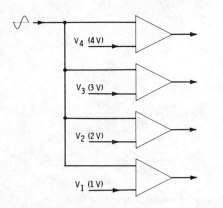

Fig. 17-4. Quantization with four comparators.

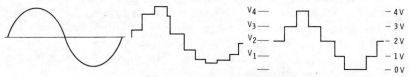

Fig. 17-5. Quantized signal.

signal is now said to be *quantized.* Note that the waveform in Fig. 17-5 does not actually appear in practice at this point; it does appear later when the digits are transformed back to an analog signal.

A more accurate waveform can be obtained by using more reference levels. The waveform using eight comparators is shown in Fig. 17-6.

Fig. 17-6. Signal quantized to eight levels.

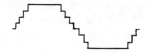

Encoding

The separate output lines from the comparators are fed into a converter circuit, where the input pattern of ones and zeros is converted into binary code. The most usual code is straight binary counting.

With four converters, the first four binary numbers would be used. With eight converters, the first eight numbers would be used. The greater the number of steps in the quantizing process, the more numbers will be needed to represent the steps. For example, if only two steps were used, then the numbers 0 and 1 would suffice. So one binary digit, or *bit,* that alternated between 0 and 1 would be the output. If four steps were used, then the binary numbers 00, 01, 10, and 11 would be enough. This is two bits that alternate between 0 and 1. For eight levels, the numbers 000 up to 111 would be needed. These numbers require three bits. The general rule that comes out of this is:

$$\text{Number of Levels} = 2^n$$

where,

n is the number of bits.

Looking back over the previous examples:

$$2 = 2^1$$
$$4 = 2^2$$
$$8 = 2^3$$

This relationship is important because it shows the number of separate lines needed out of the a/d converter to carry the digitized information. It also tells how many pulses make a word when the information is converted to serial form and sent over a line or a microwave link. This characteristic directly affects the transmission frequency and bandwidth.

In digital tv, an eight-bit word is used. This corresponds to 256 different levels in the quantized signal:

$$2^8 = 256$$

The exact circuits for converting the signal into a digital word will not be covered, because there are many ways of effecting this conversion. However, two simple methods will be shown.

Method 1—One method is basically the same as the simple process already described. The signal is fed simultaneously to a large number of comparators. A different reference level is fed to each comparator. All of the outputs feed an encoder, which produces a binary number on parallel lines or converts it to serial form on one line. For a 256-bit digitized tv signal, 255 comparators are necessary, and an eight-bit word is produced.

The main advantage of this method is that it is very fast and very accurate, which is ideal for video. Its main disadvantages are its large size and high cost.

Method 2—A second method is a very simple a/d method that is used in some digital voltmeters. Fig. 17-7 shows the basic setup.

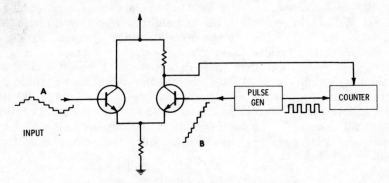

Fig. 17-7. Simple method of a/d conversion.

The step waveform is fed as input A to one side of a differential amplifier. The other input, B, is a waveform made of many very small steps, which are generated in a pulse circuit that also puts out a pulse for each step. These pulses are fed to a counter, and the count is displayed.

When input B equals input A, the differential amplifier changes state and stops the counter. The binary number in the counter is a digital representation of the original signal. Waveform B is now reset to zero, and it starts again to climb and generate numbers for the next part of the signal.

This method can be made very accurate by increasing the number of steps. If the original waveform is fed directly into the differential amplifier, the system will still work, but the samples will be at uneven time intervals.

The main drawback to this method is that it is very slow because waveform B is reset to zero each time.

DIGITAL-TO-ANALOG CONVERSION

When a string of binary numbers is received at the end of a transmission line or is played back from a digital recorder, it must be converted back to the original waveform. For this purpose, a digital-to-analog (d/a) converter is used. In most cases, this device transforms the binary number into the quantized step waveform, which is then filtered to produce the original waveform.

The most common d/a converter is the resistive ladder. There are two versions of this, shown in Figs. 17-8 and 17-9. The method shown in Fig. 17-8 is easiest to understand. Each bit has the same voltage level, so it is fed to one end of a resistor that has a value

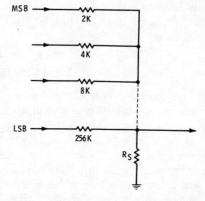

Fig. 17-8. Ladder with weighted
resistance values.

inversely proportional to the position the bit occupies in the binary number. The most significant bit (MSB) is fed to a low-value resistor; the next bit is fed to a resistor of double that value; the resistor value is doubled again for the next bit, and so on. The least significant bit (LSB) is fed to the resistor of highest value. As a result, the currents through the resistors and into the summing re-

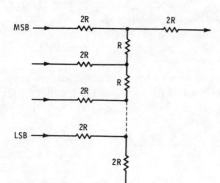

Fig. 17-9. Diagram of R-2R ladder.

sistor are proportional to the fraction of the total voltage that each bit contributes to the final signal. The final signal is developed across the low-value resistance (R_s) that sums the currents.

One of the main disadvantages of this arrangement is that precision resistors are needed. Another problem is that the very high resistance values needed for the LSB cause trouble at high bit rates.

The method shown in Fig. 17-9 is basically the same, but only two values of resistor are needed. Hence, it is easier to implement this method.

If the incoming bits are in serial form, they must first be converted to parallel form before they are applied to the resistor ladder. As the bits change, the combined current changes to produce a stepped change in the voltage output, thus re-creating the quantized step waveform.

SOURCES OF ERROR

The digital process works quite well, but it does contain a few minor problems. When a sample is taken from the original signal, it can only approximate the exact value. If the actual sample taken falls between two reference levels, the sample is rounded off at the lower level (Fig. 17-10A). This also happens when a sample crosses a reference level (Fig. 17-10B), because the sloping original waveform is truncated to a pulse of exact height.

The result of these errors is that when the quantized waveform is converted back to an analog waveform, it may be somewhat different

(A) *Between levels.*   (B) *Crossing a level.*

Fig. 17-10. Sampling errors.

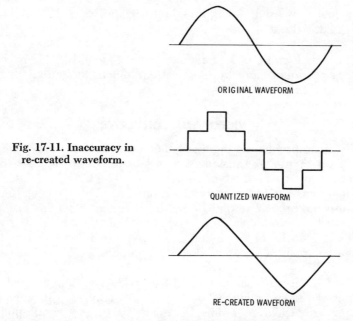

ORIGINAL WAVEFORM

Fig. 17-11. Inaccuracy in re-created waveform.

QUANTIZED WAVEFORM

RE-CREATED WAVEFORM

from the original. Fig. 17-11 illustrates this effect. The only way to prevent severe distortion of this sort is to increase the number of samples taken, make them narrower, and increase the number of reference levels used.

A major practical decision in a/d and d/a conversion involves how many samples and how many reference levels will be needed to reproduce an analog signal with acceptable accuracy. The figures chosen differ from one application to another.

THE VIDEO SIGNAL

The two main parameters for digitizing the video signal were chosen after extensive subjective viewing tests. They are:

1. Number of sample levels: The number of levels chosen is 256. This gives a reconstituted video signal that is indistinguishable from the original. Since $256 = 2^8$, an eight-bit digital number is used for the binary information. Some industrial and educational TBCs use only six bits, and some high-quality broadcast units use nine bits.

2. Number of samples taken: Since the color must be reproduced accurately, Nyquist's theorem demands that the sampling frequency be more than twice the color subcarrier frequency. In practice a sampling frequency 3 or 4 times the subcarrier fre-

quency is used, giving sampling frequencies of 10.7 MHz and 14.3 MHz.

The actual width of the sampling pulse is around 20 nanoseconds. This is governed more by practical equipment limitations than by theoretical considerations.

VIDEO A/D CONVERTER

The main problem in digitizing the video signal is the type of a/d converter to use. A satisfactory compromise between high speed, accuracy, and cost is the type shown in Fig. 17-12.

The signal is fed to a four-bit a/d converter. This contains 15 comparators, and the 15 output lines are converted to a four-bit digital word or binary number. This word represents the four most significant bits (MSBs) of the digital number. These are now fed to a d/a converter to form an approximate brightness signal with 16 steps between black and peak white.

The 16-step signal is subtracted from a delayed input signal. The output of the subtractor is an error signal only $\frac{1}{16}$ the amplitude of the original. This is amplified to exactly 16 times its amplitude and fed into another circuit with 15 comparators. The output of this circuit is converted to four bits, which are the four least significant bits of the final signal.

As a result of these processes, there are eight output lines with all the necessary bits for a signal divided, or quantized, into 256 levels. This method uses only 30 comparators instead of 255. It is often called a *subranging converter*.

APPLICATIONS

Digitizing the video signal is an important development in tv, and it is changing the character of studio equipment tremendously. The following paragraphs mention briefly some of the products that use these techniques.

Important news items from around the world are likely to originate from tv systems that are different from the NTSC system. Also, many entertainment programs produced in one country find great sales in others. Hence, converting from one tv system to another is most important. The methods used for this have been complex, expensive, and bulky, or they have been avoided by using film. The appearance of the digital standards converter has made international exchange of tv tapes much easier, cheaper, and faster.

A digital frame store digitizes and stores a complete frame of video. The advantage of this is that extremely large timing errors

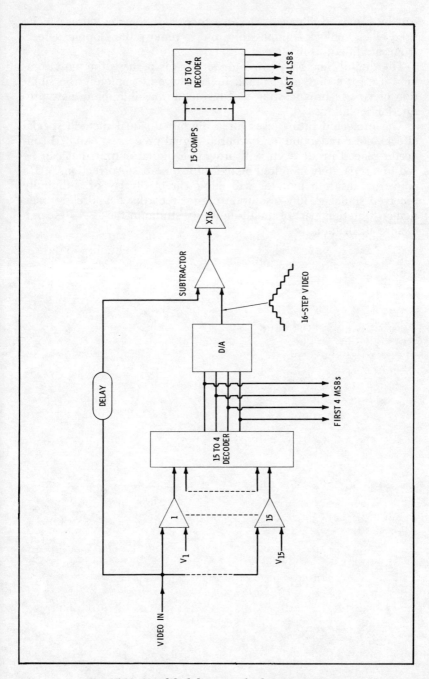

Fig. 17-12. Simplified diagram of subranging converter.

393

can be removed. Typical uses are to permit remote sources to be used when genlock is impossible, and to remove the Doppler effects from satellite transmissions.

The digital time-base corrector for vtr's is perhaps the most common use for a digital tv signal. It has altered completely the situation of vtr's in broadcasting and has made possible many new production techniques.

The projected future uses for a digital tv signal include special effects, noise reduction, and accurate signal processing. An a/d converter placed inside a vtr will provide a digital output that can be fed to a separately mounted memory for time-base correction. A d/a converter inside a monitor will allow direct display of a digitally received signal. A low-resolution digital recorder has already been demonstrated as part of an all-digital tv studio in the IBA Research Centre in Britain.

Index

A

Acoustic wave device, 271
A/d
    conversion, 384
    converter, video, 392
Afc loop, 253
Agc, 114
Alpha wrap, 59-60
Ampex
    format, 321-322
    VPR-1, 326-336
Amplifier
    input, 114
    record, 123-126
Analog
    signal, digitizing, 383-384
    -to-digital conversion, 384-389
Apc circuit, 251-253
Assemble mode, 229
Assembly editing, 220-221
AST, 329
Audio, 223-224
    playback, 300
    record, 300
    tape recorders, 23-25
    tracks, 322-324
Auto framing, 230
Automatic scan tracking, 329-331
Azimuth error, 48

B

Balanced modulator, 255
BCN machines, 315-320
Bearings, 89
Belt drive, 105-106

B-H curve, 16
Bias
    frequency of, 19
    recording, 15-19
Bidirex dial, 343-344
Bit, 387
Bosch-Fernseh BCN machines, 315-320
Brake on feed spool, 100-101
Braking, eddy-current, 158-159
Broadcast
    helical
        format, 320-321
        machines, portable, 381-382
        vtr's, nonsegmented, 320-325
    vtr
        formats, 305-325
        requirements, 305-325
Burst
    gate, 256-257
    -track method, 241-242

C

Camera, portable, 366-368
    recording from, 373-374
Capstan
    assembly, 68, 97-99
    motor, 105
        drive, 187-191
    phase servo control, 179-182
    servo, 174-182, 193, 334-336, 356
    Sony
        AV 3650, 176-182
        EV 320F, 174-176
Carrier frequencies, 30

Cassette(s), 284-285
  machines, 283-284
    mechanics of, 285-289
    portable, 380-381
    smaller, 301-304
    U-matic, 253
CCD, 271
Charge-coupled device, 271
Circumference, drum, 83
Clamp, sync, 114
Clip, white and dark, 115-116
Color
  correction, 250-253
    direct record, 239-244
    review, 259-260
  editing, 264-265
  edits in small helical machines, 265
  framing in NTSC, 264
  noise canceller, 258-259
  noncoherent, 260
  nonphased, 260
  packs, 264
  playback
    circuits, 294
    problems, 237-239
  portable, 378-380
  record circuits, 294
  test patterns, 262-264
  topics related to, 260-265
Comparator(s), 148-156, 182
  FG-type, 154-155
  output, 155-156
  servo, 195-196
Construction, mechanical, 369-370
Control(s)
  circuits, 156-157
    system, 345-346
  devices, 158-159
  panel, 336
  phasing, system, 277
  pulses, editing with, 233-234
  remote, 346
  system, 297-298
  track, 218-219, 324, 376
    playback, 146-147
    pulse, 144-147
    recording, 145-146
Conversion
  analog-to-digital, 384-389
  digital-to-analog, 389-390
Converted-subcarrier method, circuits
    for, 253-259
Converter
  a/d, video, 392
  frequency, 257-258
  standards, digital, 392

D

D/a converter, 389
Dark clip, 115-116
Dc motors, 158

Deck, 328-329
  portable, 369-371
  tape, 67-68
De-emphasis, 133-134
Degaussing, head, 91
Delay
  devices, 271
  line(s), 261-262
    electronically variable, 269
Demod-remod methods, 239-242
Demodulator, 130-133
  circuit, 331
  pim, 134-136
Detector, phase, 257
Deviation, 112
Dial, Bidirex, 343-344
Diameter, wheel, 49-50
Digital
  drive, 189
  frame store, 392-394
  standards converter, 392
  time-base corrector, 272-282
    functions, additional, 275-280
    operation of, 272-275
  -to-analog conversion, 389-390
Digitizing analog signal, 383-384
Dihedral, 90
Direct
  method, 236-237
  mode, 280-281
  record color correction, 239-244
Discriminator, FG, 350-352
Down-converted subcarrier method,
    244-250
Drive
  belt, 105-106
  rim, 106
Drivers, rotating-mechanism, 157-158
Dropout, 35
  compensation, 202-205, 275
  period, 199-201
Drum
  circumference, 83
  head, and, arrangements, 58-64

E

Echo machine, 311-313
Eddy-current braking, 158-159
Edit(s)
  color, small helical machines in, 265
  logic, 230
Editing, 348
  accessories, 230-234
  assembly, 220-221
  color, 264-265
  control pulses, with, 233-234
  erase, 216-218
  facilities, examples of, 224-230
    IVC, 228-230
    Sony
      AV 3650, 225
      EV 320F, 225-228

Editing—cont
    insert, 221-222
    servos for, 215-216
    SMPTE code, with, 231-233
Electronics, 68-74, 291-300
    playback, 71-74
    record, 71-74
    to electronics, 136-137
Encoding, 387-389
Enhancement, picture, 277-278
Equalization, 22-23
Erase
    current switching, 219-220
    editing, 216-218
    head, flying, 218
    main, 217
Error(s), 48-49
    azimuth, 48
    quadrature, 48
    sources of, 390-391
    time-base, 266
        causes of, 267-268
EVDL, 269-271

F

FG
    discriminator, 350-352
    -type comparator, 154-155
Filter, low-pass, 114
Flying erase head, 218
Fm modulator, 116-123
Format(s)
    Ampex, 321-322
    broadcast
        helical, 320-321
        vtr, 305-325
    SMPTE Type C, 322-324
        advantages of, 324
    Sony, 321
    tape, 84-86
        Ampex, 84
        EIAJ-1, 84-86
        IVC, 84
Frame store, digital, 392-394
Framing
    color, in NTSC, 264
    servo, 356-359
Frequency(ies)
    converter, 257-258
    fm, 45
    -modulated signal, 29
    range limitations, 26-29
Full-wrap machines, 63-64
    one-head, 59-62, 71-73

G

Gate, burst, 256-257
Gauge, penetration, 94-95
Guard band, 82
Guide(s), 40
    tape, 64-67

Gyro errors, 278

H

Half-wrap machines, 63-64
    two-head, 62-63, 73-74
Head(s), 15, 86-91, 107-109
    assembly, 39-43
    changing, 93-97
    connections, 92
    current, increased, 217-218
    degaussing, 91
    drive, 91
    drum, 290, 326-327
        and, arrangements, 58-64
        assembly, 77-79
        motor, 105
        recorded tracks, and, 80-83
    -motion problems, 267
    motor drive, 187-191
    output, 20
        maxima and minima in, 20
    penetration, 37
    position, 69
    servo, 161-172, 191, 331-334, 353-
        356
        Ampex 7500, 165-166
        Shibaden SV 700, 169
        Sony
            AV 3600, 168-169
            AV 3650, 172
            EV 320 Series, 166-168
    switching, 35, 47-48, 74, 201-202
    -tach pulse, 141-144
    -to-tape speed, 81
    wear, 49-50
Helical
    format, broadcast, 320-321
    machines, portable, broadcast, 381-
        382
    time-base errors, 271-272
    vtr's
        broadcast
            nonsegmented, 320-325
            requirements for, 307-309
            fundamentals of, 51-56
Heterodyne
    method, 242-244
    option, 348
    signals, 277
Heterodyning, 260-261
High-band frequencies, 45
Horizontal-sync pulse alignment, 81-
    82
Hybrid helical vtr, 309-320
Hysteresis loop, 16

I

Input signals, 141-148
Insert
    editing, 221-222
    mode, 229

Integrated circuits, 161
Integrator, 155-156
IVC 9000 Series machines, 313-315

L

Laminations, 14
Layout, physical, 347-348
Levers, 107
Limiters, 130
Linelock, 193
Lines, delay, 261-262
Linkages, mechanical, 107
Longitudinal machine, 32
  problems with, 33
Losses, 21-22
Low-band frequencies, 45

M

Machine(s), 325-363
  Sony, 336-363
Maintenance, 346-347
MDA, 182, 331-334
Mechanical
  linkages, 107
  splicing, 235
Meter(s), 209-211
  tracking, 210-211
  video, 209-210
Method adopted, 29-34
Mistracking, 69
Modulator, 116-123
  balanced, 255
  circuit, 331
  phase, 352-353
  pim, 121-123
Motor(s), 103-105
  capstan, 105
  control of feed spool, 103
  dc, 158
  drives, head and capstan, 187-191
  head-drum, 105
Multivibrators, use of in servo circuits,
  160-161

N

Noise canceller, color, 258-259
Nonsegmented broadcast helical vtr's,
  320-325

O

Omega wrap, 60
One-head machine, 128-129
  full-wrap, 59-62, 71-73
"1.5-head system," 61
Operation of machine, 300
Oscillator(s),
  scanning-erase, 229
  sine-wave, 116-119

Oscillator(s)—cont
  square wave, 119-120
  voltage-controlled, 255

P

PAL, 262
Penetration, head, 37
Phase
  detector, 257
  modulator, 352-353
  shifting, electronic, 188
Phasing controls, system, 277
Picture enhancement, 277-278
Pilot
  converted subcarrier with, 248-250
  -tone method, 242
Pim
  demodulator, 134-136
  modulator, 121-123
Pixlock mode, 194
Plate, rotating, 89
Playback, 20-21, 246-248
  audio, 300
  circuits, color, 294
  electronics, 71-74
  mode, 46-47, 163-164, 180-182
    servos in, 193-194
  servo operation, 376
  video, 126-136, 293-294
Portable
  broadcast helical machines, 381-382
  cassette machines, 380-381
  color, 378-380
Preamplifier(s), 126-130
  switching, 130
  unswitched, 130
Pre-emphasis, 114
Process mode, 281-282
Pulse
  alignment, horizontal-sync, 81-82
  control-track, 144-147
  head-tach, 141-144
  60-Hz from power line, 147
  vertical-sync, 141

Q

Quad timing problems, 268-271
Quadrature error, 48
Quantizing, 385-387

R

Ramp
  sampling, 148-150
  slice, 151-154
Record, 245-246
  amplifier, 123-126
  audio, 300
  circuits, color, 294
  electronics, 71-74

Record—cont
  mode, 45-46, 162-163, 179-180
    normal, 229
    switching, 219-220
Recorders, tape
  audio, 23-25
  principle of, 12-13
Recording
  bias, 15-19
  from
    portable camera, 373-374
    tv receiver or other source, 374-376
  servo in, 373-376
  signal onto tape, 13-15
  video, 110-126, 292-293
Reel(s)
  servos, 336, 361-363
  tables, 106-107
  tape, 105
Remote control, 346
Rf output, 298-300
Rim drive, 106
Rotating mechanism(s), 158-159
  drivers, 157-158

                S

SAM, 271
Sampling, 385
  ramp, 148-150
Scalloping, 49
Scanning speed, 81
Scott transformer connection, 188
SECAM, 262
Segmented helical vtr, 309-320
Serial analog memory, 271
Service, 346-347
Servo(s), 34, 68-71, 294-297
  capstan, 174-182, 193, 334-336, 356
    phase, 179-182
    Sony
      AV 3650, 176-182
      EV 320F, 174-176
  circuit(s), 348-363, 371
    basic, 349-350
    multivibrators in, 160-161
  comparators, 195-196
  control, 105
  editing, for, 215-216
  framing, 356-359
  fundamentals, 138-161
  head, 161-172, 191, 331-334, 353-356
    Ampex 7500, 165-166
    Shibaden SV 700, 169
    Sony
      AV 3600, 168-169
      AV 3650, 172
      EV 320 Series, 166-168
  playback
    mode, in, 193-194

Servo(s)—cont
  playback
    operation, 376
  quad-head, 47
  recording, in, 373-376
  reel, 336, 361-363
  speed control, 177-179
  tension, 359
  types of in common use, 139-141
Shoe, 40
Shuttle modes, stable pictures in, 278-279
Skew errors, 268
Skewing, 49
Slant-track principles, 56-58
Slice, ramp, 151-154
Slow speed, 206-209
SMPTE
  code, editing with, 231-233
  Type C format, 322-324
Sony
  format, 321
  machine, 336-363
Space loss, 35-37
Speed
  control servo, 177-179
  head-to-tape, 81
  scanning, 81
  tape, 31
    linear, 82-83
Splicing, mechanical, 235
Spool
  feed
    brake on, 100-101
    motor control of, 103
  take-up, control, 101-103
Standards converter, digital, 392
"Sticktion," 31, 78-79, 80
Still frame, 206-209
Stop button, 222
Subcarrier
  converted
    method, circuits for, 253-259
    pilot, with, 248-250
  down-converted, method, 244-250
Surfaces, smooth, 80
Switching
  head, 35
  record and erase current, 219-220
Switchlock, 193
Sync clamp, 114

                T

Take-up spool control, 101-103
Tape
  deck, 67-68
  -dimension problems, 267-268
  formats, 84-86
    Ampex, 84
    EIAJ-1, 84-86
    IVC, 84

Tape—cont
  guides, 64-67
  -motion problems, 267
  path, 43-44
  recorder(s)
    audio, 23-25
    principle of, 12-13
  recording signal onto, 13-15
  speed, 31
    linear, 82-83
  tension, 35, 99-103
  timers, 345
TBC, 272, 348
  operation, 280-282
Tension
  servo, 359
  tape, 35, 99-103
Test patterns, color, 262-264
Time-base
  corrector, digital, 272-282
    functions, additional, 275-280
    operation of, 272-275
  errors, 266
    causes of, 267-268
    helical, 271-272
Timers, tape, 345
Timing problems, quad, 268-271
Tone, motor-generated, 147-148
Tonewheel mode, 193
Track(s)
  angle, 81
  audio, 322-324
  control, 218-219, 324, 376
  length, 81
  quad-head, 43
  recorded, head drum and, 80-83
  slant-, principles, 56-58
  video, widths, 82
Tracking, 69-71
  meter, 210-211
  scan, automatic, 329-331
Transformer connection, Scott, 188
Transport
  tape, 328-329
  Type
    1, 285-288
    2, 288-289

Tuner, tv, 298-300
Two-head machine, 130
  half-wrap, 62-63, 73-74
Type
  1 transport, 285-288
  2 transport, 288-289

V

Velocity compensation, 277
VERA, 10, 32-33
Vertical
  advanced, 279-280
  -sync pulse, 141
Vfo, 350
Video
  a/d converter, 392
  circuits, 340-343, 370-371
  meter, 209-210
  playback, 126-136, 293-294
  recording, 110-126, 292-293
  signal, 391-392
Voltage-controlled oscillator, 255
Vtr
  broadcast
    formats, 305-325
    requirements, 305-325
  color
    direct, 280
    heterodyne, 280-282
  helical
    fundamentals of, 51-56
    hybrid, 309-320
    nonsegmented broadcast, 320-325
    requirements for broadcast, 307-309
    segmented, 309-320

W

Wavelength, formula for, 28
Wheel diameter, 49-50
White clip, 115-116
Wrap(s), 55
  alpha, 59-60
  omega, 60